Aquaculture Landscapes

Aquaculture Landscapes explores the landscape architecture of farms, reefs, parks, and cities that are designed to entwine the lives of fish and humans.

In the twenty-first century, aquaculture's contribution to the supply of fish for human consumption exceeds that of wild-caught fish for the first time in history. Aquaculture has emerged as the fastest growing food production sector in the world, but aquaculture has agency beyond simply converting fish to food. *Aquaculture Landscapes* recovers aquaculture as a practice with a deep history of constructing extraordinary landscapes. These landscapes are characterized and enriched by multispecies interdependency, performative ecologies, collaborative practices, and aesthetic experiences between humans and fish. *Aquaculture Landscapes* presents over thirty contemporary and historical landscapes, spanning six continents, with incisive diagrams and vivid photographs. Within this expansive scope is a focus on urban aquaculture projects by leading designers— including Turenscape, James Corner Field Operations, and SCAPE—that employ mutually beneficial strategies for fish and humans to address urban coastal resiliency, wastewater management, and other contemporary urban challenges. Michael Ezban delivers a compelling account of the coalitions of fish and humans that shape the form, function, and identity of cities, and he offers a forward-thinking theorization of landscape as the preeminent medium for the design of ichthyological urbanism in the Anthropocene.

With over two hundred evocative images, including ninety original drawings by the author, *Aquaculture Landscapes* is a richly illustrated portrayal of aquaculture seen through the disciplinary lens of landscape architecture. As the first book devoted to this topic, *Aquaculture Landscapes* is an original and essential resource for landscape architects, urbanists, animal geographers, aquaculturists, and all who seek and value multispecies cohabitation of a shared public realm.

Michael Ezban is an architect, landscape designer, and scholar. His work explores landscapes and buildings designed to mediate relations between humans and other animals. Published and exhibited internationally, Ezban's writing and design focus on aquaculture landscapes, waterfowl hunting grounds, and equestrian facilities. He is a recipient of the Maeder-York Family Fellowship in Landscape Architecture from the Isabella Stewart Gardner Museum and the Charles Eliot Traveling Fellowship from the Harvard University Graduate School of Design. Ezban is an Assistant Professor in Landscape Architecture at the University of Virginia, and a cofounder of VanderGoot Ezban Studio, a research-based practice.

Aquaculture Landscapes is a breath-taking book—full of historical drawings, ingenious diagrams, and superb photography—that demonstrates how we can revolutionize our relationships with aquatic life. Grounded in multispecies urban theory and dreams of coexistence instead of exploitation, Ezban offers both concrete examples and speculative designs from around the world that will transform landscape architecture practice. His book is the first to not only argue for a post-human urbanism, but to demonstrate how landscape architects can go about creating a zoöpolitan urbanism for the future. *Aquaculture Landscapes* is a must-read for all landscape architecture students, faculty, and professionals.

—Jennifer Wolch
Dean, College of Environmental Design, UC Berkeley, USA

Michael Ezban presents an original and informative book on an extremely intriguing subject: aquaculture landscapes—fish farms as seen through the lens of contemporary landscape architecture. His work opens and invites us all into vast and exciting new territory for landscape architecture practice. Through examination of public landscapes where nutrient cycling, biological conservation, remediation, and vernacular farming practices combine to inspire a sense of wonder and pleasure, *Aquaculture Landscapes* illustrates both the art of survival and the art beyond survival.

—Kongjian Yu
Dean and Professor, Graduate School of Landscape Architecture, Peking University, China
President, Turenscape

Aquaculture Landscapes is a thoughtful, articulate, forward-thinking contribution to a deeply problematic yet still-growing sector of food production. Combining intelligent prose with illuminating illustrations, Michael Ezban presents an alluring vision for how new, ecologically and ethically sensitive systems can—indeed, must—be designed for a livable future. With what I've seen of the state of contemporary industrial aquaculture, this is a much-needed blueprint.

—Jonathan Balcombe, author of *What a Fish Knows*

We are all born of water, and our primordial relationship with aquatic ecologies endures. In this lush volume, Michael Ezban both plumbs the depths and surfaces common currents to reveal an aqueous terrain worthy of navigation for the Anthropocene. Through rigorous historical research and insightful contemporary precedents from aquaculture to angling, *Aquaculture Landscapes* offers fresh thinking and timely designs for a richer, more biodiverse world. Ezban's design research intelligently articulates how we might materially and metaphorically cohabit with our oldest of relatives—the fish—and other species beyond the human.

—Nina-Marie E. Lister
Graduate Program Director, School of Urban + Regional Planning, Ryerson University, Canada

Aquaculture Landscapes is an incredible achievement! Michael Ezban's beautiful maps, diagrams, and renderings unpack the diverse and integrated worlds that humans and fish inhabit. This book is a celebration of our ingenious and resourceful history as humans guiding and cultivating the aquatic landscape.

—James Prosek, author of *Eels*

We are ecological beings. We interact with our fellow humans as well as with other species and our environments. But our interactions with other species tend to be one-sided: we take much more than we give. In *Aquaculture Landscapes*, Michael Ezban takes compelling deep dives into cohabited landscapes from around the world, and he offers a hopeful vision for how we might rebalance and reconfigure our relationships with fish and other aquatic life through design.

—Frederick Steiner
Dean, Stuart Weitzman School of Design, University of Pennsylvania, USA

AQUACULTURE LANDSCAPES

Fish Farms and the Public Realm

Michael Ezban

Routledge
Taylor & Francis Group

LONDON AND NEW YORK

First published 2020
by Routledge
2 Park Square, Milton Park, Abingdon, Oxon OX14 4RN

and by Routledge
605 Third Avenue, New York, NY 10017

First issued in paperback 2021

Routledge is an imprint of the Taylor & Francis Group, an informa business

British Library Cataloguing-in-Publication Data
A catalogue record for this book is available from the British Library

Library of Congress Cataloging-in-Publication Data
A catalog record has been requested for this book

ISBN 13: 978-0-367-78380-8 (pbk)
ISBN 13: 978-1-138-21835-2 (hbk)

Publisher's Note
This book has been prepared from camera-ready copy provided by the author.

Typeset in Avenir.

For my father, Morris, who loved to go to the sea.

01 View to cascading raceways at Posto Aquícola Ribeiro Frio, the state-run fish farm located within Laurisilva of Madeira, Portugal, a UNESCO World Heritage Site. The centuries-old system of *levadas* (irrigation canals) supply the farm with fresh water. Rainbow Trout *(Oncorhynchus mykiss)* are farmed here to stock the levadas and other waterways of Madeira.

Contents

Foreword

Imagining Aquaculture

We have been eating fish since long before we were human.[1] Fish have been as fundamental and formative to our development as any terrestrial species. Fishing and eating fish, either raw or cooked, predate our emergence as *Homo sapiens*. Yet, fish farming is a decidedly human invention, deeply associated with our species. The terrestrial cultivation of fish for human consumption is found across cultures and climates as diverse as human habitation itself.

This fascinating book presents the astonishing diversity of these aquaculture practices and the landscapes they construct. *Aquaculture Landscapes* assembles essential knowledge on both ancient and contemporary human practices for readers across the design disciplines. The case studies curated in this volume present a compelling, comparative account of these cultural practices as well as ample evidence of the variable relations between the humans and non-humans that cohabit these landscapes. These cases offer thick readings of constructed, species-specific ecological systems, and they augment contemporary interest in the design of mutualistic relationships between species across complex webs of ecological feedback. This study is particularly timely, as ecological performance has emerged as a lens for understanding urbanism and species-driven projects in contemporary design practice.

Michael Ezban's rigorous original research illustrates not only our relationship with fish and farming, but it also describes the broad range of societal and cultural conditions that aquaculture supports. Ezban rightly reads the material practices of aquaculture and their social economies as forms of landscape. By claiming these sites as landscape, Ezban invokes the social construction and cultural complexity that the term originally invoked in its European origins, yet he extends that frame to describe constructed landscapes found across nearly every human culture. Ezban posits aquaculture as an ancient form of socially constructed landscape with significant import for contemporary design discourse on urbanism, a claim that is resonant with the current disciplinary interest in agricultural production and its implications for landscape architecture and the shape of the city.

Ezban's reading of farms, reefs, hatcheries, and associated aquaria as socially constructed landscapes is reinforced through his attentiveness to the complexity with which multiple species are ordered, in various relationships to one another, at these sites. His evocative account of these multispecies entanglements draws on contemporary posthuman discourse that examines the status and meaning of non-humans in the social construction of culture. Ezban's drawings, diagrams, and texts conspire to produce an exemplary portrayal of aquaculture landscapes, and the species they assemble, as extraordinary cultural constructs. This book has much to teach us about human relations with the non-human world, as well as the roles of landscape in mediating those relations and imagining that world.

Charles Waldheim, Harvard University, USA

01 Louis-Joseph Yperman, *La Pêche au Vivier*, raised wall mural, Palais des Papes, Avignon, France, 1910. Copy after original from 1343-1344.

Acknowledgements

This project on aquaculture landscapes has been six years in development. Before it was a book, it was a series of travels and experiences with people, fish, and landscapes—an embodied practice of research that I initiated to get a feel for what aquaculture is, what it has been, and what it can be. My journeys included visits to extant sixteenth-century carp ponds in the Czech Republic; the 36 ft diameter fiberglass tanks in a nondescript warehouse in Massachusetts that house thousands of barramundi, a fish native to Australian billabongs; interdependent and intertwined Andalusian salt and fish farms; the ruins of a *piscina* on the Italian coast where a Roman emperor and his guests would relax alongside eels; and the alluvial forest of a nineteenth-century hatchery landscape in France. In my travels I savored the taste of lightly seasoned Common Carp; I spied egrets stalking Northern Pike from a bird blind; I joined six men who harvested a group of Grey Mullets from a drained pond with a seine and then witnessed the slow death of those fish on ice; I tossed feed pellets to thrashing Rainbow Trout in concrete raceways; I observed sinuous European Eels swimming in a coastal lagoon and watched their long bodies being threaded on skewers and fired; and I sketched anglers effortlessly pulling Brown Trout out of a tailwater on a Tuesday morning.

My experiences were enriched by conversations with farmers, hatchery managers, and biologists who provided me with insight into the world of fish culture. My deepest thanks to J. Miguel Medialdea, the Quality and Environment manager of the Veta la Palma fish farm in Spain, for walking with me across the productive landscape that he has stewarded for many years. Witnessing, experiencing, and discussing the intricate workings of that constructed ecology was crucially formative to my thinking and made me feel hopeful for a future of ecological aquaculture. My forays into aquaculture landscapes across Europe and North America would not have been possible without financial support from the Charles Eliot Traveling Fellowship from the Harvard University Graduate School of Design, a prize I am truly grateful to have received.

For all my travels, there were many landscapes I did not visit. In these instances, I am so thankful for the scholars who agreed to dig back into their slide libraries to unearth photographs from their own time spent at aquaculture landscapes. Thank you, Peter Edwards, Luohui Liang, Kenneth Ruddle, and Clark Erickson for your willingness to share your beautiful images with me.

The Isabella Stewart Gardner Museum in Boston was a wonderful place to live and work for three months in 2014 as I explored these topics. I am grateful to museum director Anne Hawley for conversation and perspective on how my work resonates with the museums' previous artists in residence. Thanks to all members of the museum staff who offered comments, questions, and support as I shared my work, and especially to JoAnn Robinson for extending generous hospitality throughout my stay. I am so appreciative that the Maeder-York Family Fellowship Selection Committee, including Anita Berrizbeitia, Julia Czerniak, Teresa Galí-Izard, Charles Waldheim, and Richard Weller, saw potential in my research proposal. Special thanks to Alan Berger, who proposed that I consider shifting focus from the Boston Harbor to the Quabbin Reservoir and the waterways that form the metropolitan water system; I'm so glad I took your advice!

I am indebted to Charles Waldheim, my mentor since 2011, who has helped me hone this project and shaped my larger sense of landscape architecture as well. I am so grateful for our many conversations in recent years on angling and waterfowl hunting over meals and in front of audiences—such unexpected topics! Thank you, Charles, for your thoughts at the front of this book, they mean so much to me. I am also so appreciative of the guidance, encouragement, and perspectives from several of the faculty at the Harvard University Graduate School of Design during my time in the Master in Landscape program, especially Pierre Belanger, Chris Reed, John Dixon Hunt, and Mark Laird.

I appreciate the contributions that various designers, colleagues, research assistants, and editors brought to this book. Thank you to Siena Scarff for the thoughtfulness and creativity you conveyed to the cover design; you created an evocative threshold for the book. I can't thank my colleague David Bayer enough for bringing rigor, dedication, and patience to the project of beautifully rendering the fifteen case study transects in this book. Thank you, David, for the many hours, and for always being up for one more round of edits. I am also grateful for my research assistants who traveled far to capture aquaculture landscapes through photography, especially Jason Manongdo who exhibited excellent drone-piloting skills in Hawaii and Kentucky, and Boyu Li for her beautiful photographs and travel in China. Finally, I appreciate the guidance of Amy Johnston and Hannah Ferguson at Routledge, and I have been so very well-served by the editorial acumen of Jake Starmer, who went extra rounds with my near-final drafts.

Exhibitions and public lectures were wonderful opportunities to share and advance this work during its development. I am grateful to Brian Kelly and Ronit Eisenbach for the gracious invitation to mount an exhibition of this work at the Kibel Gallery at the University of Maryland School of Architecture, Planning & Preservation in 2016. I am also grateful to David Hill and Rob Holmes for extending an offer to exhibit this work and lecture at Auburn University School of Architecture, Planning and Landscape Architecture in 2017. I appreciate the opportunity to share and discuss my aquaculture research along with other emerging landscape academics at the Landscape Dialogues Symposium at the University of Pennsylvania School of Design in 2017. Finally, I am grateful for the invitation to deliver a public lecture at the Isabella Stewart Gardner Museum in 2014, a first opportunity for me to draw the lines that connect my work on aquaculture landscapes, waterfowl hunting grounds, and equestrian facilities that has spanned my career.

A special thank you to my colleagues at the Virginia Tech Washington-Alexandria Architecture Center, especially Laurel McSherry, Paul Kelsh, and Nathan Heavers for inviting me to collaborate in an environment and community that I found stimulating and formative. I appreciate the trust, guidance, support, and friendship you all have offered me over the past several years.

Finally, Jana, I could not have done this without your help and your love. I am so grateful that you lent your discerning eye to the project, and I truly appreciate that you offered just the right edits at just the right time. Thank you for believing in me. I love that my kids were drawing alongside me over the last few years; thank you Peter, for your vivid and creative drawings of fish and farms, and Anna, for your loose and imaginative explorations. Thank you Luiza, for the unending love you have offered in so many ways, and Al, for slaughtering commas, reshaping sentences, and other helpful edits. And finally, thank you Henry and Mary VanderGoot, for your generosity in always being willing to engage and support my creative works.

01 The diversity of historical and contemporary fish farms across the globe has produced a rich, transcultural heritage of aquaculture landscapes.

Introduction

Aquacultures

Most fish never left the primordial waters. They stayed submerged in marine and freshwater environments and evolved over millions of years alongside molluscs and crustaceans. They diversified into tens of thousands of species, weathered waves of extinction and proliferation, and developed an extraordinary range of behaviors, migration strategies, specializations, and morphologies. And the fish that left the sea? Consensus among paleontologists, evolutionary biologists, and marine scientists affirms evidence that some primeval fish emerged on land about 375 million years ago[1], and that these pioneers from the Age of Fishes are the ancestors of almost all vertebrates alive today, including humans.[2]

Paleontologist Neil Shubin suggests that our human bodies offer clues about the origins we share with fish. He describes our "inner fish," noting the resemblance in the bone structure of human arms and hands to ancient fins and the similarities between the organization of our heads and those of our aquatic forebearers.[3] Building on the long lineage of work in evolutionary biology, paleontology, and genetics, Shubin's anatomical observations add dimension to the relationship between humans and fish.

The first material evidence of encounters between humans and fish include the recent discovery in northern Kenya of catfish bones marked with cuts from stone tools. This find dates early hominin consumption of fish to nearly two million years ago and factors into discussions of how fish provided human ancestors with essential nutrients for expanding their brain size.[4] The earliest record of *Homo sapiens'* systematic harvesting of shellfish dates back 164,000 years and was discovered in a coastal cave in South Africa.[5] Around 25,000 years ago in ancient France, a human carved the earliest known representation of a fish into limestone. The fish is depicted in such remarkable detail that experts identified it as a spawning Atlantic Salmon (*Salmo salar*).[6]

Only in the past several millennia did humans begin to construct landscapes for the capture and culture of fish. With the advent of this practice, fish flourished in new homes that had been built precisely with them in mind, and material evidence reveals that humans thrived with fish in these landscapes.

The oldest known examples date to the Neolithic era. The Gunditjmara people at Lake Condah in south-western Victoria, Australia, constructed an extensive system of interconnected traps, ponds and habitats for Shortfin Eels (*Anguilla australis*). These have been dated to 6,600 years ago.[7] Excavated into a rugged lava-flow terrain, this landscape includes 35 km of channels that once steered migrating eels into traps. The traps were connected to a constructed wetland eel habitat that increased the fish's availability to humans beyond the migration season. The management of the eel habitat as a component of this landscape has led some scholars to consider the Lake Condah site to be one of the first and largest eel farms.[8] Archaeologists also believe they have found the remains of numerous circular stone huts near the excavated eel channels. In addition to the huts, interspersed hollowed trees, which are thought to have been used for smoking and cooking eels, serve as probable traces of a culinary culture.[9]

A remarkable diversity of terrestrial fish farms has emerged across the globe since the Neolithic era. These farms form a rich, transcultural heritage of *aquaculture landscapes*. Aquaculture landscapes shape, and are shaped by, a range of encounters between fish and humans. The production of these landscapes is informed by the variable relations between the species that cohabit them, as well as the dynamic economies and cultures in which they are embedded. Aquaculture landscapes are geographically located sites that are subject to shifts in climate, species populations, and migration patterns. These landscapes range in scale from ponds to watersheds, and their management has regional and global ecological and economic implications.

Examples of aquaculture landscapes are found in China, where estuaries and mountainsides alike have been transformed through the construction of ponds and terraces to generate polycultures of carp, rice, vegetables, and fruit (see Case Studies 10 and 11). In the twenty-first century, China produces two-thirds of the world's freshwater aquaculture output, and a detailed treatise on carp culture published in China in 475 BCE, speaks to the longevity of the country's aquaculture traditions. In the pre-Columbian savanna of Baures, Bolivia, thousands of kilometers of weirs were constructed by the people of Baure to channel seasonal floods and turn the savanna into a fishery (see Case Study 13). In the Třeboň Basin, Czech Republic, a regional-scale system of fishponds, dams, and canals that was constructed in the sixteenth century continues to produce tremendous quantities of carp for European markets. The landscape also provides recreational opportunities, flood protection, and biodiverse habitat (see Case Study 01).

Exemplary aquaculture landscapes can be characterized as sites where fish have agency and freedom to express species-specific behaviors in biodiverse habitats, novel aesthetic experiences and recreational activities are enabled, colocated programming and infrastructural functions produce diversified economies, and mutualistic relations between humans, fish and plants are constructed. As such, aquaculture landscapes have significant cultural and ecological importance, and they are also particularly relevant to contemporary discourses in landscape architecture and related design disciplines.

Intensive Fish Monocultures

Despite the deep history of wide-ranging aquaculture practices and the multifunctional landscapes they produce, aquaculture in the twenty-first century is largely framed and practiced in narrow terms. The Food and Agriculture Organization (FAO) of the United Nations, for example, defines aquaculture simply as the farming of aquatic organisms, including vertebrates, invertebrates, and plants. According to the FAO, aquaculture involves both an intervention to enhance production of fish, through feeding or breeding practices for instance, as well as ownership of the stock being cultivated.[10]

The growth of global aquaculture surged in the last fifty years. In 2019, the production of fish for human consumption is the fastest growing food production sector in the world.[11] Growth has been fueled by "intensive" aquaculture, high-density monocultures that rely on industrially manufactured feed pellets and hormonally induced breeding. In many cases, these fish feedlots have supplanted more resilient ecological practices.[12] They also reduced the range of activities that traditionally took place in aquaculture landscapes to the singular program of transforming fish into food.

In addition to the production of fish for human consumption, aquaculture also produces fish to stock waters for recreational fishing and to restore populations of imperiled species. In service of this function, the US Federal and State fish hatcheries have for nearly 150 years enacted the controversial practice of introducing non-native fish species, such as Rainbow Trout (*Oncorhynchus mykiss*), into rivers and streams across the country (see Part Three). In 2004, hatcheries in the United States stocked 1.7 billion fish, a population that constitutes the largest total weight of fish stocked in the country in any given year, since adequate records have been kept.[13]

These modes of fish farming have an intellectual precursor in the French aquaculture revolution of the nineteenth century. That era saw the construction of state-operated hatcheries, such as the *piscifactoire* (fish factory) in France, to produce salmonoids for stocking blighted rivers to satisfy the growing interest in recreational fishing among elites (see Case Study 06). Technologies and knowledge, along with fish and eggs, were exported to countries eager to institute their own programs.

Characterized by environmental historian Darin Kinsey as efforts to manipulate aquatic species to fit human cultural expectations, the French aquaculture revolution was linked to the contemporaneous Western ideology of improving nature, as well as acclimatization practices, such as the global distribution of non-native species, that emerged during the period of European colonialism.[14] Kinsey writes that policies and programs of the revolution "promoted the imperial conquest of water, and its diffusion of aquatic species had profound global consequences."[15] The legacy of the nineteenth-century revolution continues to reverberate in contemporary intensive aquaculture, where anthropocentric and productivist narratives frame fish farming in terms of yields, and species are genetically modified for economic gain. These paradigms and practices are exemplified in the contemporary farming of Atlantic Salmon, the same species of fish carved in limestone by human hands in ancient France.

The intensive farming of Atlantic Salmon began roughly forty years ago. In 2014, this species was globally the most economically valuable cultured fish and in 2016, more than 99% of all salmon consumed globally were farmed.[16] Intensive farming of Atlantic Salmon is characterized by high densities of fish, penned in open water. These salmon are prone to parasites and disease, and the application of vaccines and antibiotics is routine.

The weight gain of wild salmon is tied to their migratory patterns. Their growth surges prior to their summer spawning run, and slows in winter. Farmed salmon, on the other hand, are genetically modified to be hungry and gain weight year-round.[17] Pelleted feed that satisfies the intensified growth pattern of farmed salmon is produced from wild stocks of fish. Around 2.5 kg of forage fish are required for every kilogram a farmed salmon gains.[18] Salmon void most of the nutrients from the consumed pellets, creating nutrient overloads and eutrophication in the water near the farms.[19]

Another issue is that salmon that escape from farms differ from native populations of wild salmon in terms of their genetics, behavior, and physiology. There are concerns that the mingling of farmed and wild salmon will lead to reduced productivity of the overall salmon population and decreased resiliency of wild salmon.[20] Despite the many negative impacts, intensive practices such as Atlantic Salmon monocultures persist. As the rising global human population becomes more urban and fish consumption increases, the expectation is that intensive aquaculture practices will continue to fuel the growth of global aquaculture.[21] In 2014, a long-expected but astonishing milestone was met. For the first time in history, the contribution of aquaculture to the supply of fish for human consumption exceeded that of wild-caught fish.[22]

Landscapes for Multispecies Coalitions

It is in the context of these trends and histories that *Aquaculture Landscapes* delves deeper into theories and practices of mutualism, polyculture, species interdependency, and urban cohabitation, offering alternatives to intensive monocultures in an attempt to realign aquaculture with landscape and humans with fish.

This introductory essay discusses aquaculture from three disciplinary perspectives: 1) landscape architecture (the primary lens through which this book is written), 2) human-animal studies, and 3) science of aquaculture and ecology. One recurrent theme found in all three disciplines concerns "multispecies coalitions," alliances between humans and non-humans that inspire collaborative patterns of interaction between species, challenge the typical power relations found in conventional animal agriculture, and enable animal agency in the coproduction of shared landscapes.

Contemporary practitioners and academics in the field of landscape architecture have shown leadership by envisioning, theorizing, and enacting progressive forms of aquaculture. These works illustrate many roles that landscape can play in the contemporary, cohabited public realm. Landscape architect Kate Orff, whose widely-recognized work includes a multifunctional oyster and human habitat in New York Harbor in the United States, urges her audience to think "beyond a built environment conceived exclusively for human consumption and comfort" in order to "address the wider global ecosystem as a shared space for all species."[23]

The field of human-animal studies includes geographers, anthropologists, ethnographers, and scientists who explore more-than-human perspectives and places of multispecies entanglements. Works in this field are premised on an understanding of relations between humans and non-humans as simultaneously biological, cultural, economic, ethical, geographical, and political.[24] Ethologist Jonathan Balcombe, for example, examines breakthroughs in sociobiology, neurobiology, and other fields that explore fish sentience, as well as their diverse social and emotional lives.[25] Balcombe illuminates a perspective of fish as beings with intrinsic value, and he calls for their inclusion in what bioethicist Peter Singer calls the "expanding circle of moral concern."[26]

A regrounding of aquaculture in the wisdom of sustainable indigenous practices is advocated in many quarters of the marine and freshwater sciences. Prominent aquaculture scholar Barry Costa-Pierce argues, "Aquaculture must become less short-term and less production oriented, and become more ecologically, community, and culturally based."[27] The FAO has also attempted to encourage more "ecological approaches" to aquaculture in their literature and programs.[28]

Aquaculture Landscapes joins the forward-thinking, multidisciplinary discourse outlined above. This book brings together many examples of farms, reefs, parks, aquaria, and cities, where the lives of fish and humans are inexorably entwined. The aquaculture landscapes in this book are charged spaces that enable vivid encounters, socialization, and collaboration between species. The stakeholders, stewards, and fish at these sites form a constellation of relationships that underlie the social construction of aquaculture landscapes. These relations are foundational in making twenty-first century aquaculture into a phenomenon characterized by practices, attitudes, and landscapes that can only be cocreated by multispecies coalitions.

Geographers Jody Emel, Connie Johnston, and Elisabeth Stoddard suggest that we "hold ourselves accountable to [animals] in changing, exploratory, respectful, and

collectively oriented relationships—relationships rife with contradictions, vulnerabilities, violence, and uncertainties."[29] The landscapes examined in this book enable such relationships in part because they resist easy categorization and embody contradictions. They are habitats, but they are not wild. They yield less fish than intensive practices, but they provide a range of invaluable ecosystem services that can be difficult to quantify. They are spaces of power imbalances between species, yet they can also be environments where fish have agency. They offer humans opportunities to marvel, empathize, and collaborate with fish while at the same time bringing an end to their lives through harvest. In short, the aquaculture landscapes described here are thorny territory. They are open-ended experiments in the social construction of a cohabited public realm.

Aquacultures in the Discipline of Landscape Architecture

Western traditions and philosophies of landscape architecture have origins in the design of landscape parks at seventeenth- and eighteenth-century English estates, where multispecies cohabitation was integral to function and aesthetics. Actively managed fishponds figured in these landscapes, joining deer parks, waterfowl decoys, pheasantries, livestock paddocks, dovecotes, and rabbit warrens. The management of animals and their habitat was directly tied to the varied economies and recreational programs of the estate, and management practices informed the evolving aesthetics of the landscape park.[30] As landscape historian Tom Williamson notes, "These diverse activities were not simply fitted round or hidden away from the dominant aesthetic. They lay at the very heart of the landscape park . . . forming the very essence of its structure."[31]

Prior to the seventeenth century, the construction of fishponds by the Norman aristocracy in England was common. Fishponds remained as one of the principal status features at estates in Medieval England. Two types of fishponds were typically constructed at medieval estates: *vivaria*, which were large breeding ponds constructed by damming and flooding low ground, and *servatoria*, which were small holding ponds adjacent to the residence for temporary storage of fish transferred from the vivaria. Freshwater fish were still considered status objects in the early eighteenth century when formal garden design had come into fashion. As a result, the ponds of the medieval era were often adapted to include ornamental fountains, cascades, and defined sidewalls. Garden archaeologist Christopher Currie finds that estate accounts from the era indicate that despite aesthetic modifications, the aquaculture function of these ponds was still treated "very seriously."[32]

Dyrham Park in Avon is a prime example of such a transformation. Five medieval fishponds at the estate underwent radical aesthetic transformation and were incorporated into an elaborate formal garden in 1704. Following these changes, the ponds continued to be used for rearing and breeding fish. An account from 1710 of the fish kept in Dyrham Park's ponds includes notes on aquaculture practices with projections of up to ten years out for when fish would be ready for harvest.[33] Observing trout, perch, and carp at Dyrham Park garden, the theorist Stephen Switzer noted that the "scaley residents" were not disturbed by the frequent visitation of people.[34]

A more pastoral landscape aesthetic was pursued by landscape designers, gardeners, and homeowners later in eighteenth century England. The design of water features shifted from formal geometries toward the "naturalized" shapes of ponds that followed existing contours of the land. Currie notes the resonance in both shape and construction between these informal ponds and the vivaria fish breeding ponds of the

medieval period.[35] As farming of carp and perch for the table fell out of favor at English estates in the eighteenth century, the new picturesque ponds supported a burgeoning interest in the sport of angling among elites. It is the aesthetics and philosophies of pastoral landscape design—which, in England, were infused with histories of multispecies cohabitation, aquaculture, and angling—that influential landscape architects like Fredrick Law Olmsted sought for inspiration while developing public landscapes and parks in the United States, in the nineteenth century.

Landscape architects in the twenty-first century are once again designing for multispecies cohabitation. A system-based understanding of ecology, advanced by biologist Eugene Odum in the twentieth century, informs contemporary urban design theory.[36] Charles Waldheim, James Corner, Kongjian Yu, and other influential thinkers in the field of landscape architecture theorize urbanism as dynamic and interconnected social and biophysical systems where landscape is both a model and medium for urbanism.[37] Landscape architect Kate Orff poses important questions for the project of designing urban ecologies for a multispecies public realm: Can the notion of working for "clients" also include attending to "perceived human and animal needs?"[38] Can "new mutualistic systems be introduced that encourage the success of both human and non-human inhabitants?"[39]

Contemporary landscape architects affirm these possibilities through the design of infrastructural systems and landscapes that function as habitats. Latz+Partner created a significant new habitat for migratory neotropical birds at the once barren Hiriya Landfill in Israel, by designing a lush *wadi* at the foot of the landfill and a vegetated "oasis" at the top. In the Netherlands, West 8 transformed the landscape of Amsterdam Airport Schiphol into a birch forest and pollinator habitat, making it among the first projects in a contemporary global movement toward airport apiaries. Designers create migratory wildlife crossings like the ARC Wildlife Bridge in Colorado, United States, by Michael Van Valkenburgh Associates, which was designed to span an interstate highway with striated woodland ecologies and provide safe crossings for black bear, elk, and other species.

Increasingly, designers are also crossing fish habitats and farms with infrastructure and public space to create multifunctional aquaculture landscapes. In urban contexts, these landscapes employ mutually beneficial strategies for fish and humans to address coastal resiliency, revitalization of urban ecologies, municipal waste and stormwater management, and other pressing urban challenges that are central to contemporary landscape practice. For example, as part of the larger redesign of the Seattle waterfront by James Corner Field Operations, the Elliott Bay seawall is remade as a biodiverse salmon migration corridor and public promenade. Another example is found in China, where Turenscape's design for Yunhe Park in Yichang transforms a derelict fish farm into a stormwater management park and urban wildlife habitat (see Part One).

Landscape theorist Elizabeth Meyer has enriched the discourse on infrastructural landscapes as habitat by probing their aesthetic dimensions. She posits that human experiences of emotion, reflection, and even pleasure are intensified through "prolonged, vivid, and strange encounters with constructed nature," and these aesthetic experiences differ from those that arise in the wild.[40] As part of her argument, Meyer suggests that "numerous, simultaneous individual experiences in a public space comprise an aesthetic collectivity and create new ways of living in and thinking about the environment."[41] Extending upon Meyer's thinking, and Orff's queries, other questions arise. If urban

infrastructure is designed to serve as loci of encounters between species in the city, can our intensified, decentering experiences at these "constructed natures" foster recognition of sentient fish and other animals as contributors to the "numerous, simultaneous, individual experiences" that comprise collectivity? Can granting fish and other animals subjectivity, recognizing our kinship with them, and planning for their success and ours, seed the emergence of "new ways of living" in multispecies societies?

More-Than-Human Perspectives on Aquaculture and Angling

Contemporary discourse in the field of human-animal studies has been fueled by, among other things, the theoretical shift to a more-than-human framework that recognizes animals as social subjects and actors in the world.[42] Work in this field, and the allied field of multispecies ethnography, examines animal agency, collaboration and interdependency between species, and the implications of cohabitation. These works provide useful frameworks to explore varied relations between fish and humans at aquaculture landscapes.

One of the themes of a more-than-human perspective is that embodied and recurrent experiences with non-humans are vehicles for decentering and connecting to their worlds. Urbanist and geographer Jennifer Wolch writes that cohabitation with other animals provides the "local, situated, everyday knowledge of animal life required to grasp animal standpoints or ways of being in the world, to interact with them accordingly in particular contexts, and to motivate political action necessary to protect their autonomy as subjects and their life spaces."[43] Geographer Catherine Johnston, building on anthropologist Tim Ingold's philosophy of dwelling, advocates "day-to-day living and working" relationships with other animals as "a way of knowing about and knowing with animals."[44] She notes that these experiences are "by their nature noisier, smellier, messier and, in some cases, bloodier than we might like to think," but they open paths to empathic relationships with non-humans.[45]

In "Livelier Livelihoods: Animal and Human Collaboration on the Farm," Emel, Johnston, and Stoddard advocate animal agriculture characterized by "flatter hierarchies," where farming practices are aligned with animals' known behavioral tendencies to create "a partnership type of on-farm relationship."[46] While the authors explore pigs in permaculture systems, their paradigms can be applied to reconsider humans and fish in aquaculture as "co-workers." Humans provide food, shelter, and healthcare while fish "work as foragers, fertilizers, seed spreaders, mothers, caregivers, and food providers."[47] Ecofeminist scholar Donna Haraway, exploring the ethics of animal labor, writes, "To be in relation of use to each other is not the definition of unfreedom and violation."[48] Framing and constructing farms as habitat, where fish have agency and can make choices relative to feeding, socializing, and exploring, is one form of acknowledging that fish are "working subjects, not just worked objects."[49]

Angling landscapes are also sites where embodied experiences foster complex connections with non-human worlds. Through fieldwork conducted in rivers in Yorkshire, United Kingdom, geographers Christopher Bear and Sally Eden find that anglers "do not see the cold blood or scaly bodies of fish as alien or as a barrier to attempting to understand and, to an extent, empathize with them."[50] The geographers describe what they see as the everyday attempts of anglers "to think like fish."[51] Anglers adapt their behavior to share the rhythms of fish communities, as well as individual

fish. Likewise, fish seem to adapt to the practices of anglers, by changing where, when, and what they feed on over time. Thus, for Bear and Eden, angling is a coproduced and reciprocal activity—a "transformative practice whereby anglers and fish adapt through their coconstitutive encounters."[52] New streams built specifically for both spawning and angling exemplify that the complex, coconstitutive encounters between fish and humans also produce landscapes (see Case Study 14).

At both the farm and the stream, humans kill fish. The ethical implications of killing fish for food and sport are contentious and debate on the subject is perennial. Haraway observes that the act of killing can itself create opportunities for mindful reflection. She writes that killing is an inevitable act, since "eating means also killing, directly or indirectly," but she adds that "killing well is an obligation akin to eating well."[53] Humans, she writes, should resist killing with a sense of self-certainty or ethical resolution. We should "learn to live responsibly within the multiplicitous necessity and labor of killing," always, "yearning for the capacity to respond and recognize with response, always with reasons but knowing there will never be sufficient reason."[54]

Day to day encounters with fish, working alongside fish at farms, thinking like them in streams, and mindfully killing them, are some of the decentering activities enabled at aquaculture landscapes. Embodied experiences like these can foster empathic responses to fish, which in turn can fuel critical examinations of the practices, coalitions, and ethics that shape aquaculture and angling in a more-than-human world.

Implementing Ecological Approaches to Aquaculture

Recent years have seen calls for ecological and system-based approaches to sustainable fish farming by institutions like the FAO as well as communities of aquaculture scholars. The FAO convened a discussion with international aquaculture experts in 2007 to explore an "ecosystem approach to aquaculture" (EAA) that the group defined as "a strategy for the integration of [aquaculture] within the wider ecosystem in such a way that it promotes sustainable development, equity, and resilience of interlinked social and ecological systems."[55] In 2018, a critical review of EAAs effectiveness in joining aquaculture with local socio-economic development found that this strategy had been impactful.[56]

The Veta la Palma fish farm in Spain is a remarkable example of an EAA. The farm is a public-private partnership in which a commercial fishery leases land from the Doñana Nature Reserve and adheres to strict environmental regulations (see Case Study 03). At the farm, constructed wetland basins are flooded with water from the Guadalquivir River and the fish raised in this polyculture environment freely forage for shrimp and microalgae that proliferate in the basins. The wetlands cleanse the water contaminated from upland agriculture—the farm's discharge is cleaner than the water that enters it. Over 250 species of migratory birds routinely visit the site, and visiting birders augment the farm's revenues.

An EAA can also inform urban aquaculture that is multifunctional and linked to social networks. In the edited volume *Urban Aquaculture*, scholar Costa-Pierce positions urban aquaculture as "a vital, functional, and necessary component of urban ecosystems."[57] He recommends redirecting flows of municipal wastewater toward aquaculture wetlands to fuel fish production and vitalize habitats in the city. An extraordinary model for such a system is the East Kolkata Wetlands in India (see Case Study 09). Other aquaculture scholars describe the need to cultivate social and aesthetic experiences around aquaculture in urban ecosystems, to educate and inspire the next

generation of urban aquaculturists. They point to successes in integrating aquaculture in urban school curricula as well as visceral community events like clam bakes and fish fries.

Costa-Pierce advocates that we encourage the "art of aquaculture in urban areas and plan creatively for its beauty and utility in revitalized cities."[58] He and others who call for urban aquaculture with ecological infrastructural, social, and aesthetic dimensions also recommend an integration of aquaculture and urban planning processes as a means to achieve it. Such a shift could increase collaboration between aquaculturists and landscape architects—a natural partnership, given landscape architecture's disciplinary expertise in envisioning and designing multifunctional, multispecies public infrastructure.

Toward Entwining

Aquaculture Landscapes gathers contemporary discourses, historical accounts, and evocative visual representations, and views this assemblage through the disciplinary lens of landscape architecture. This approach informs a critical reading of aquaculture as landscape. This book recovers aquaculture as a practice with a deep history of constructing extraordinary landscapes, and posits landscape as the preeminent medium for the future integration of aquaculture and urbanism. In so doing, *Aquaculture Landscapes* aspires to support and expand the roles that landscape architects play in assembling the multispecies coalitions that coshape aquacultures for resilient coastlines, biodiverse urban ecologies, and a cohabited public realm.

In 2019, the *Catalog of Fishes*, the authoritative global reference for taxonomic fish names, lists over 35,000 known fish species.[59] Of those thousands, only six hundred species have ever been farmed.[60] The relatively small number of fish species selected for cultivation belies the likely trillions of individual fish that humans have encountered in aquaculture landscapes. For millennia, a lineage of landscapes shaped a myriad of experiences—and fluid, ever-changing relationships—between humans and these countless sentient beings. Our common histories bind humans and fish together. These bonds form a basis from which we can imagine and build the shared landscapes that further entwine our lives.

Organization of the Book

Aquaculture Landscapes features essays, case studies, and representations that are organized into three parts.

Part One: Situating Aquaculture Landscapes explores projects by leading designers working at the nexus of aquaculture and urbanism. Six projects in Asia, North America, and Europe are situated relative to historical and contemporary aquaculture practices as well as contemporary landscape design and theory.

Part Two: Surveying Aquaculture Landscapes presents fifteen aquaculture landscape case studies that describe a broad range of forms and practices of aquaculture located across five continents and ten countries. Each case study features key data, written descriptions, images, and a range of original analytical drawings.

Part Three: Depicting Aquaculture Landscapes is an image essay that places evocative contemporary drawings, models, maps, and paintings into a long, transcultural tradition of aquaculture landscape representation. These depictions of real and imagined aquaculture landscapes reveal drawing and modeling as ongoing, fertile processes through which humans discover relationships to farmed fish.

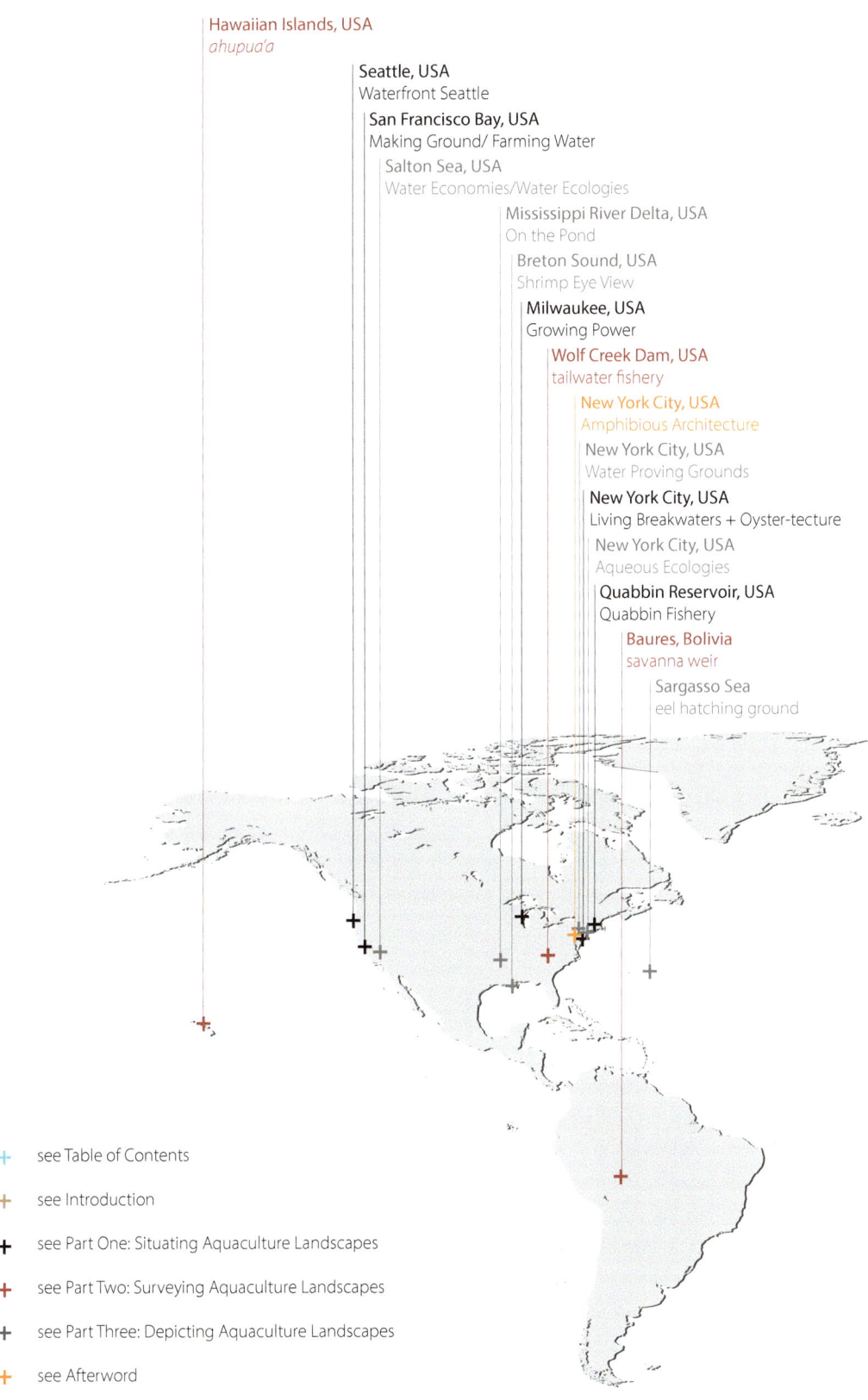

Hawaiian Islands, USA
ahupua'a

Seattle, USA
Waterfront Seattle

San Francisco Bay, USA
Making Ground/ Farming Water

Salton Sea, USA
Water Economies/Water Ecologies

Mississippi River Delta, USA
On the Pond

Breton Sound, USA
Shrimp Eye View

Milwaukee, USA
Growing Power

Wolf Creek Dam, USA
tailwater fishery

New York City, USA
Amphibious Architecture

New York City, USA
Water Proving Grounds

New York City, USA
Living Breakwaters + Oyster-tecture

New York City, USA
Aqueous Ecologies

Quabbin Reservoir, USA
Quabbin Fishery

Baures, Bolivia
savanna weir

Sargasso Sea
eel hatching ground

+ see Table of Contents

+ see Introduction

+ see Part One: Situating Aquaculture Landscapes

+ see Part Two: Surveying Aquaculture Landscapes

+ see Part Three: Depicting Aquaculture Landscapes

+ see Afterword

02 Geographic distribution of constructed and speculative landscapes described in *Aquaculture Landscapes*. Landscapes from six continents and from sixteen countries are discussed.

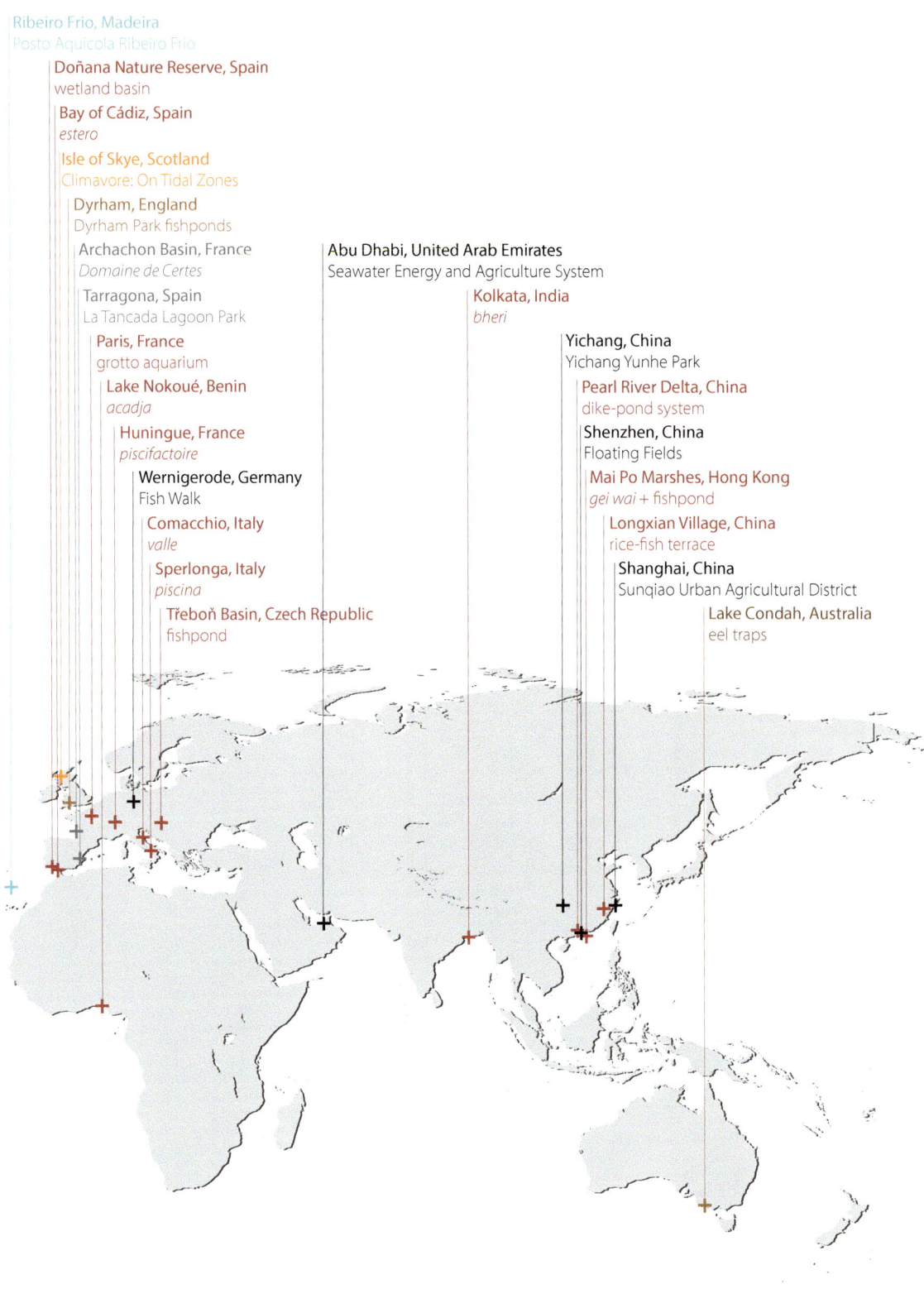

Ribeiro Frio, Madeira
Posto Aquícola Ribeiro Frio

Doñana Nature Reserve, Spain
wetland basin

Bay of Cádiz, Spain
estero

Isle of Skye, Scotland
Climavore: On Tidal Zones

Dyrham, England
Dyrham Park fishponds

Archachon Basin, France
Domaine de Certes

Tarragona, Spain
La Tancada Lagoon Park

Paris, France
grotto aquarium

Lake Nokoué, Benin
acadja

Huningue, France
piscifactoire

Wernigerode, Germany
Fish Walk

Comacchio, Italy
valle

Sperlonga, Italy
piscina

Třeboň Basin, Czech Republic
fishpond

Abu Dhabi, United Arab Emirates
Seawater Energy and Agriculture System

Kolkata, India
bheri

Yichang, China
Yichang Yunhe Park

Pearl River Delta, China
dike-pond system

Shenzhen, China
Floating Fields

Mai Po Marshes, Hong Kong
gei wai + fishpond

Longxian Village, China
rice-fish terrace

Shanghai, China
Sunqiao Urban Agricultural District

Lake Condah, Australia
eel traps

Situating Aquaculture Landscapes

01 SCAPE, *Oyster-tecture*, 2009. Detail of a cross section of the proposed Palisades Reef in New York City Harbor. The constructed reef is designed as a multifunctional public space that provides habitat for oysters, mussels, and eelgrass, dissipates wave energy, and creates opportunities for water-based recreation.

Designing Ichthyological Urbanism

Landscape architects, teamed with forward-thinking municipalities, ecologists, marine biologists, citizens, and ecosystem engineers like oysters and salmon, imaginatively design aquaculture landscapes for resilient and biodiverse twenty-first-century cities. These coalitions and their creative works are critically informed by contemporary design discourse on performative aquaculture ecologies that address urban challenges, such as sea-level rise and sustainable urban food systems; the theorization of urban animal agency in shaping the form and identity of cities; research into relations between animals' bodies, behaviors, and urbanism; and the historical precedents of urban aquaculture.

Among the most well-known contemporary urban aquaculture landscape projects is Oyster-tecture, a proposal for multifunctional oyster habitat in New York Harbor. The project was developed by the landscape architecture firm SCAPE, and commissioned by the Museum of Modern Art in 2009. Oyster-tecture features an extensive, off-shore field of nets that are stretched between pylons—an armature for a reef ecology consisting of oysters, mussels, and eelgrass. The molluscs filter and improve local water quality as they grow, and the reef attenuates strong waves that threaten the waterfront. An array of small docks set within the system enable multiple recreational activities.[1] Oyster-tecture's form recalls historical oyster-farming practices such as the Bouchot system, a wooden stake-net system for cultivating mussels that was deployed over many hectares in the Anse de l'Aiguillon in France beginning in the thirteenth century.[2]

The Sunqiao Urban Agricultural District (SUAD) in Shanghai, China, by the US-based firm Sasaki, is another recent proposal for integrating aquaculture ecologies and urbanism. SUAD is a living laboratory for innovative, urban food production practices. Aquaculture that is woven into various spaces and systems of the district enables daily public engagement with fish. Biodiverse, glassed-in office courtyards feature interior aquaponic ponds that are fed by stormwater and greywater harvested from rooftops and plumbing fixtures. And an open-air aquaculture park that mixes aquaponics, floating greenhouses, and algae farms, functions as a district-scale water-filtration system. At SUAD, fish make possible the goal of outputting cleansed water into the Huangpu River.[3]

Jennifer Wolch, a leader in the theorization of animal agency in processes of urbanization, calls for a "zoöpolis," a new "political ecology of people and animals in the city."[4] She casts animals as influential actors that affect the form, ecologies, and economies of cities; they are "critical to the making of place and landscape."[5] As evidence, she cites the outsize roles of salmon in the social and material construction of Seattle, United States. In Seattle, salmon drive economies, they galvanize citizen activism, their migrations informed the recent redesign of the waterfront, and they are afforded special protections—the entire metropolitan area is deemed a critical salmon habitat.[6]

Theories and recognition of animals' diverse roles in urban environments are augmented by empirical research into reciprocal relations between animal genetics, behavior, and urbanism. Animals that inhabit cities not only experience rapid change in their genetic makeups, but catfish, birds and other animals also exhibit remarkable behavioral adaptations to feed and survive in cities.[7] Conversely, studies also explore how animal bodies and behaviors affect changes to cities. Research into fish sentience and the social worlds they construct, for instance, informs ongoing discourse on animal rights and welfare,[8] which shapes policy on issues such as the creation of urban marine habitat. Moreover, animal behaviors with remediative and infrastructural applications, such as the ability of oysters to filter urban contaminants, are increasingly harnessed. Contemporary designers operate in the context of our growing knowledge of complex interdependencies and reciprocity between cities and the species that inhabit them.

In addition to theories and emerging research on multispecies urbanism, historical precedents of integrated urban aquaculture can also inform contemporary approaches. Wastewater-fed aquaculture systems are prevalent in Asia and across sub-tropical zones. At the East Kolkata Wetlands in India, the various species of fish that forage in low-density habitat basins are integral to the functioning of constructed wetlands that treat municipal wastewater, provide food and commerce for underserved populations, and enable a range of ecosystem services for the city (see Case Study 09).

Urban aquaria have long served as sites for impactful public interface with fish. While they are problematic in their displacement of fish, aquaria also fuel the imagination and curiosity, inspiring good stewardship of fish habitats. The atmospheric Aquarium du Trocadéro, a nineteenth-century grotto aquarium and garden in the heart of Paris, served for decades as a landscape for Parisians to marvel at displays of fish, bred to restock rivers blighted by French industrialization. The aquarium cooperated with hundreds of French angling and aquaculture societies and offered public courses to aspiring aquaculturists on the emerging art of fish farming (see Case Study 07).

Aquaculture also drives processes of informal urbanism. The lakeside village of Ganvié, Benin, is populated by approximately 20,000 Tofino people as well as thousands of Blackchin Tilapia (*Sarotherodon melanotheron*). The village is a dynamic field of constructed fish habitat, small islands, agriculture plots, and elevated houses. The hundreds of tree branches planted in the shallow water attract tilapia, eventually accumulate silt, and then are colonized by vegetation. Through this gradual process, fish habitat transitions into islands for crop farming. At Ganvié, the culture of fish simultaneously feeds and forms an evolving village (see Case Study 08).

The essays to follow describe exemplary contemporary landscape works at the nexus of aquaculture and urbanism. The essays situate these works within contemporary landscape theory, and are arranged relative to the themes of *resiliency*, *polyculture*, and *adaptive reuse*.

Resilient Aquaculture Coastlines examines designs for resilient urban waterfronts in North America that integrate fish and human communities. The reefs, shellmounds, and seawalls described in this section demonstrate that urban flood-resiliency strategies can help restore populations of salmon, oysters, and other species critical to their ecosystems.

Catalytic Polycultures explores how constructed aquaculture ecologies can have a transformative effect on urban sites and systems. Among the projects discussed is a public aquaponic ecology in China, a country with a deep history of polyculture. This section also features a trout hatchery in the United States reimagined as a commercial aquaponic system and habitat for self-sustaining populations of spawning trout.

Post-Aquaculture Adaptations examines the adaptive reuse of former fish farms as public landscapes that are informed and enriched by legacies of aquaculture. Projects in China and Europe illustrate the adaptation of ponds, dikes, and hydrologic flows of derelict farms as key components of urban development strategies.

02

03

04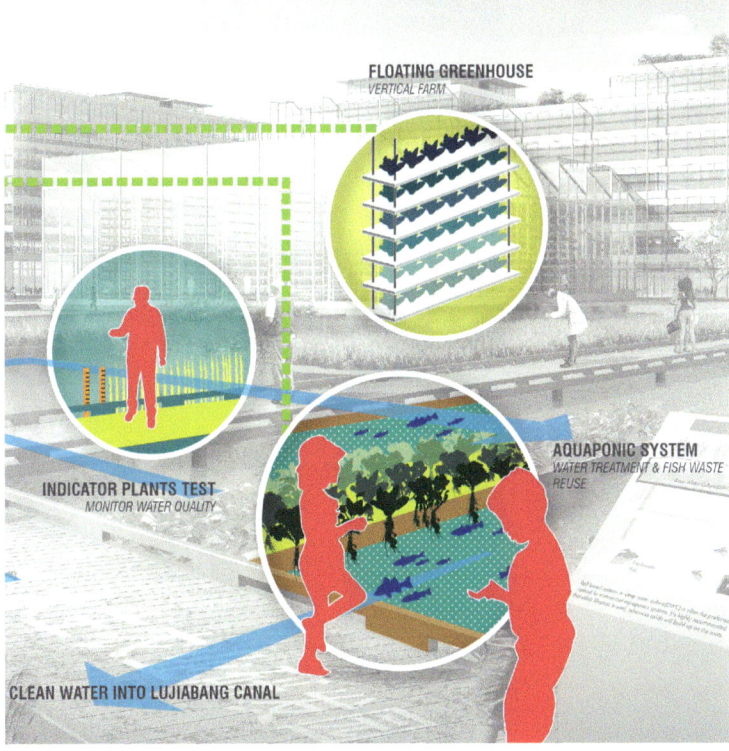

02 Sasaki, *Sunqiao Urban Agricultural District*, 2016. Aerial oblique of proposed innovation district in Shanghai, China.
03 *Sunqiao Urban Agricultural District*. Detail of perspective rendering at outdoor aquaponic park.
04 *Sunqiao Urban Agricultural District*. Detail of park diagram illustrating aquaponics linked to district water cleansing.

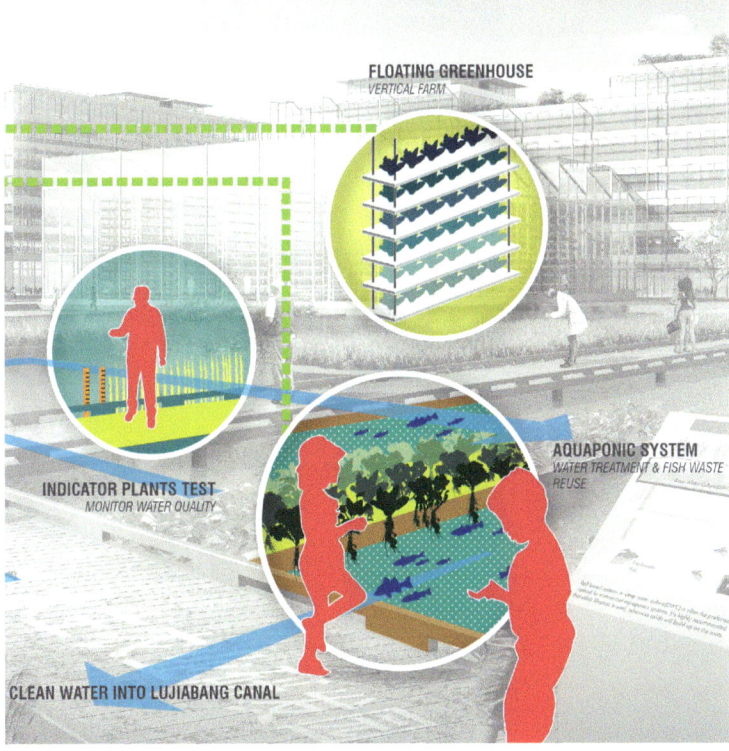

05 Tom Leader/TLS, *Making Ground/Farming Water*, 2010. Plan depicting proposed aquaculture and agriculture plots that infill between new urban mounds at Corte Madre in San Francisco Bay.

Resilient Aquaculture Coastlines

Monofunctional flood-protection infrastructure, such as the seawalls that are typical of urban coastlines, are prone to failure and deleterious to the marine ecologies that existed before the infrastructure was built.[9] Reconceptualizing and remaking this infrastructure into aquatic habitats, farms, and migration corridors has become part of the design vocabulary of contemporary landscape architects, who propose adaptive and safe-to-fail strategies for resilient urban coastlines. Resiliency, the notion that infrastructure, landscapes, and ecosystems can withstand and recover from a range of disturbances such as flooding, has emerged as a guiding principle for urban waterfront design. Ecologist Nina-Marie Lister suggests, "If resilience is to be a useful concept in informing design strategies, it must ultimately instruct how to change safely—to adapt, with transformative capacity—rather than to resist change by relying on the illusion of a perpetual normal." What Lister describes can be thought of as living infrastructure.[10]

Constructed oyster reefs are one type of adaptive, living infrastructure at the nexus of aquaculture and coastal resiliency. At the coastline of the Gulf of Mexico in the United States, the Nature Conservancy has been working with oysters—the ecosystem engineers of estuaries and bays—to construct new reefs. These reefs will protect coasts from erosion and storm surges, filter water, and increase the habitat of many marine species, including economically important commercial and sport fish.[11] Underscoring the value of oyster reefs to municipalities that struggle to fund obsolete and aging infrastructure, recent studies of the adaptive capacity of reefs in North American estuaries reveal that the speed at which reefs grow can outpace anticipated sea-level rise.[12]

Living Breakwaters, a project by SCAPE that was one of the winning proposals for the United States Department of Housing and Urban Development's Rebuild by Design Initiative, envisions breakwaters and shallow coastal edges as adaptive ecological infrastructure, reducing flood risk in areas severely damaged by Hurricane Sandy in 2012.[13] The project proposes long chains of breakwaters along Staten Island's coast, designed with perpendicular "reef streets" that create biodiverse habitats for juvenile fish and oysters (see Part Three). Landscape theorist Christophe Girot notes the transformative capacity of this project, writing, "It is the visionary scale at which an entire coastal topology will be changed to become an agent of environmental resilience that marks a real paradigm shift in landscape practice."[14] Living Breakwaters is also notable for the degree to which it builds social resiliency by fostering educated and empowered publics that are engaged in the stewardship of this living infrastructure.[15]

The two projects that follow demonstrate the design of flood-resilient urbanism underpinned by aquaculture and aquatic habitat. Making Ground/Farming Water, by Tom Leader Studio, is a radical proposition to reshape the topography of an urban estuary in San Francisco Bay. A second project, Waterfront Seattle, by James Corner Field Operations, reimagines a conventional seawall as a dynamic waterfront landscape for fish and humans.

Making Ground/Farming Water, by Tom Leader Studio

In 2009, the San Francisco Bay Conservation and Development Commission initiated a competition titled, "Rising Tides," and invited designers to develop coastal resiliency strategies to address projected sea-level rise in San Francisco Bay, California. The entry by Tom Leader Studio (TLS), titled, "Making Ground/Farming Water," reinvents the topography in the low-lying urban estuary of Corte Madera Creek. TLS envisions a series of multifunctional mounds that reference historical modes of inhabitation in the region. The mounds are surrounded by an intertidal, flexible mat of aquaculture and agriculture. This strategy structures open-ended processes of flood-resilient urban development that could be incorporated at other urban estuaries and low-laying zones across the Bay Area.[16]

Initial phases of the process proposed by TLS include the construction of a cluster of mounds in the areas soon to be inundated by water. Low-rise, tilt-up-concrete structures, roadways, and crushed stone become the raw materials for this new archipelago. Ovular mounds, ranging in scale from 20 to 60 ha would be created through a process of layering and shaping the debris. The crests of these landforms, which peak above the projected height of sea-level increases, become sites for neighborhoods, energy parks, commercial districts, and reserves. The mounds are imagined by TLS as being connected to each other by elevated roadways.

The area between mounds is a productive zone for aquatic and terrestrial farming and marshes within the tidal zone. Primary-navigation channels, along with smaller lateral canals, allow farm-boat access to productive plots. Inundated plots are sites for "scaffold reefs," which are frameworks for oyster cultivation. They are constructed with recycled sewer and plumbing pipes, filled out with reused electrical conduits, and then driven into the mud or supported by anchored pontoons. These flexible frameworks can be extended in response to tidal zones moving upland and are protected from wave energy by sub-tidal constructed oyster reefs. Upland zones between mounds are constructed with soils, topping a base of debris and structured into an array of farming plots with cross-cutting canals. This plot-and-canal system is reminiscent of the Aztec agriculture system of *chinampas* in central Mexico.

The mounds proposed by TLS resonate with massive, centuries-old middens that combined shellfish consumption and land making that were once prevalent across the Bay Area. The Ohlone Native American people of the Northern California coast settled the area and began constructing hundreds of shellmounds starting around 3,000 BCE. These shellmounds, which ranged in size from 9 m to nearly 200 m in diameter and rose to 9 m tall, typically had ovular footprints.[17] They were constructed of millions of mollusc shells, particularly the shells of the Blue Mussel (*Mytilus edulis*) and the Bent-nosed Clam (*Macoma nasuta*).[18]

The shellmounds are thought to have been constructed as individual mounds and in clusters and located in estuaries or other waterfront zones where native shellfish were populous. Archaeologist Kent Lightfoot notes that these "imposing sites once dominated the cultural landscape of the greater bay system."[19] Today, archaeologists interpret the uses of the shellmounds and their roles in three ways: as simple refuse dumps, as specialized cemeteries and monuments for burial rituals, or as multifunctional mounded villages.[20] The largest shellmounds were decimated by urban and agricultural development in the twentieth century, and only a few have been protected from further damage in designated park areas.

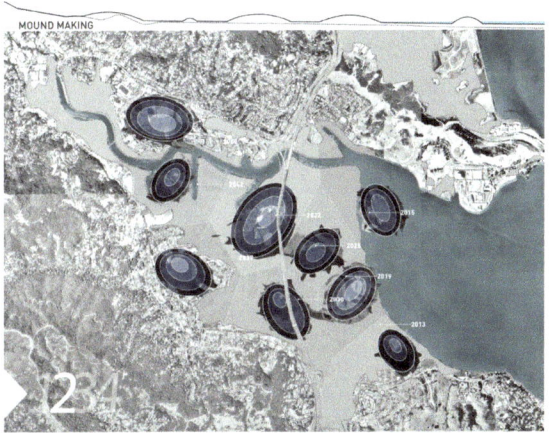

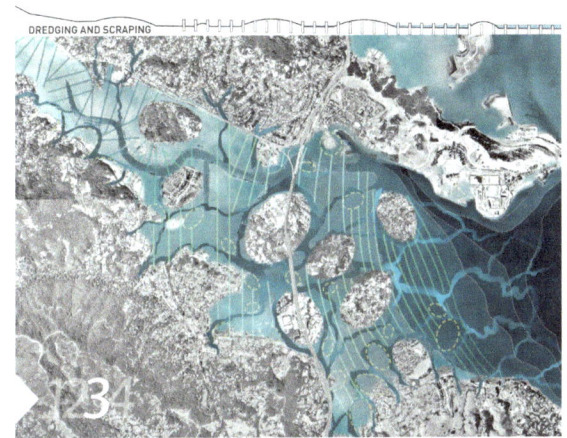

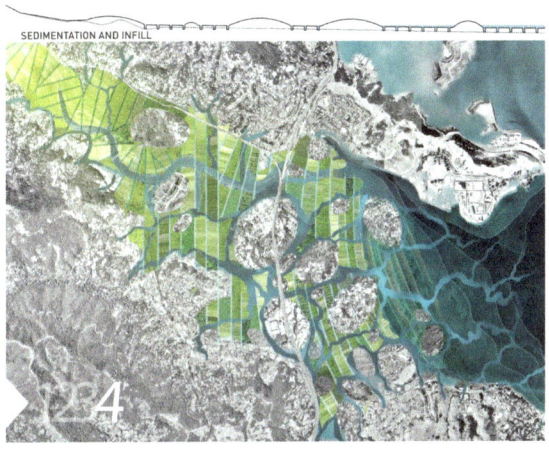

URBAN DEVELOPMENT

PARK ZONE
ENRICHED SOIL CAP
ROAD
EROSION CONTROL MATRIX

CONSOLIDATED STABILIZED DREDGE

UPPER CLAY LENS

CONSOLIDATED RUBBLE FINES

CONSOLIDATED RUBBLE

RECYCLED CRUSHED BASE ROCK

CURB SECTION FROM REDWOOD HWY
FLOOR SLAB FROM PUBLIC STORAGE
TILT-UP SLAB FROM APPLE STORE

ASPHALT RUBBLE FROM MACY'S
PARKING LOT
LOWER CLAY LENS
20TH CENTURY ARTIFICIAL FILL
18TH CENTURY BAY MUD

12TH CENTURY CONSOLIDATED
BAY MUD

LATE HOLOCENE ALLUVIUM

HOLOCENE ALLUVIUM
OVER BEDROCK

06 Tom Leader/TLS, *Making Ground/Farming Water*, 2009-10. Site plan depicting emergence sequence and key processes.
07 *Making Ground/Farming Water*. Mound section reveals urban development atop layers of detritus and geology.
08 *Making Ground/Farming Water*. Rendering of oyster farming with a mound in background.

Nels Nelson, a pioneering archaeologist who conducted initial surveys and excavations of the shellmounds in the Bay Area in the early twentieth century, promoted the idea that the Ohlone people constructed villages on the mounds they made. Some mounds may have been occupied for thousands of years, with each generation living atop the remains of previous ones. Nelson estimated that at any one time there may have been up to 12,000 people living on these shellmounds across the Bay Area, each contributing to the ongoing process of land-making through the shucking of mussels and clams.[21]

In imagining the "making of ground" for the Corte Madera of San Francisco Bay, TLS explicitly references the historical processes of topographic transformation undertaken by the Ohlone people. TLS's proposal to create mounds from urban rubble, fish farms from repurposed plumbing pipes, and topsoil from sediment, creates continuity with the Ohlone people's mode of coastal inhabitation, which was to live on and with the deposition of shells and detritus at the water's edge.

Examination of the strata of the ground and understanding the anthropological and natural processes at play in the making of ground, are central and recurrent themes in Leader's work.[22] Landscape architect Philipe Coignet notes Leader's "compulsive need to understand what is beneath the ground level and his desire to visualize the construction of a landscape since its origins—to retrace the evolution of each component and its interaction with others."[23] In Making Ground/Farming Water, TLS proposes an urban substrate that not only reconnects the Bay Area to its ecological and anthropological history, but also rebuilds a relationship between humans, fish, and land. This approach expands the contemporary discourse on flood resiliency, so that it extends beyond performance and protection, to include historical resonance and memory.

Waterfront Seattle, by James Corner Field Operations

In 2012, the City of Seattle, Washington, in the United States, approved plans to redesign the city's waterfront as an urban ecology to benefit fish and humans. James Corner Field Operations (JCFO) lead a multidisciplinary team that designed Waterfront Seattle, a flood-resilient waterfront landscape, "located within a region of transition between two ecological communities, the aquatic communities of Elliott Bay and the upland communities in the urban neighborhoods abutting the waterfront."[24] JCFO organized various zones along the urban waterfront into a range of aquatic and upland habitats, including kelp forests, cobble reefs, gravel beaches, intertidal benches, prairie, and maple alder woods.

Central to the multi-year project proposal was the redesign of the deteriorating seawall that was constructed as flood-protection infrastructure in the early twentieth century. The seawall replaced the mudflats and sloping beaches of the intertidal zone of Elliott Bay. The seawall negatively affects the behavior of the wild and artificially propagated populations of Pacific salmon as they move through Elliott Bay during their spring migration from the Duwamish River to the Pacific Ocean because it lacks the surface texture needed to promote biodiversity.[25] The shipping piers adjoining the wall also greatly reduce light penetration into the aquatic environment. The inhospitable edge created by the seawall infrastructure forces migrating salmon into deeper, well-lit waters, and this puts the salmon at risk as food for predators. Among the species of fish affected by the seawall is the Chinook Salmon (Oncorhynchus tshawytscha), a fish that is listed as threatened under the Endangered Species Act of 1973.[26]

09

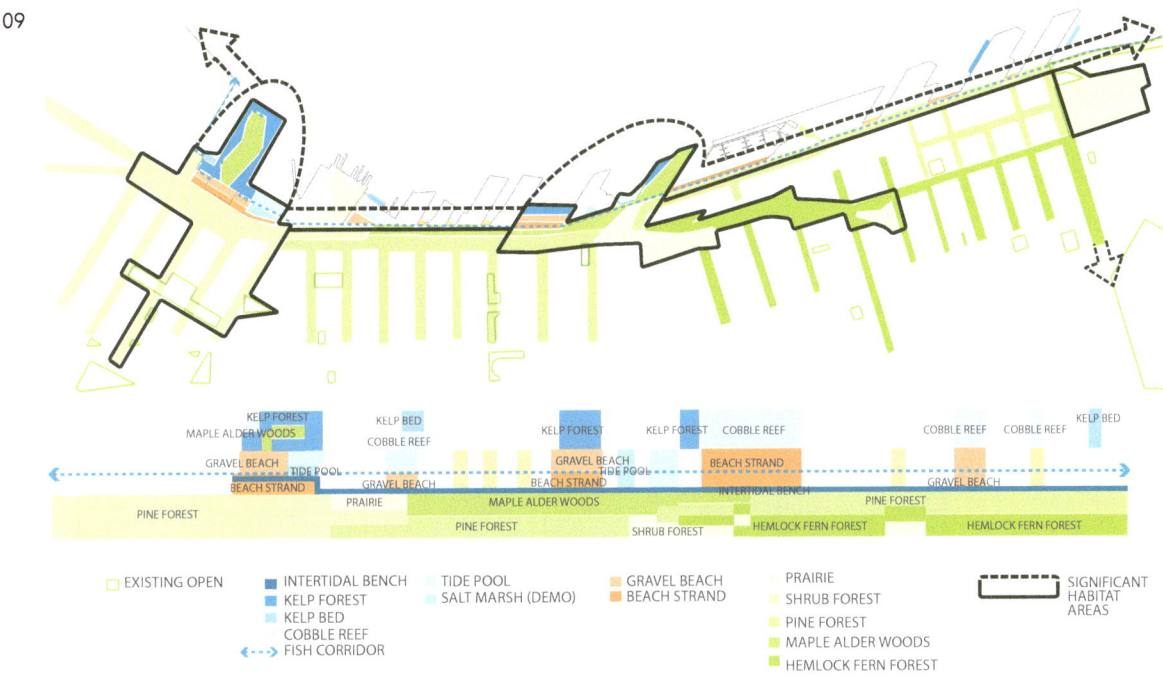

EXISTING OPEN   INTERTIDAL BENCH   TIDE POOL   GRAVEL BEACH   PRAIRIE   SIGNIFICANT HABITAT AREAS
KELP FOREST   SALT MARSH (DEMO)   BEACH STRAND   SHRUB FOREST
KELP BED   PINE FOREST
COBBLE REEF   MAPLE ALDER WOODS
FISH CORRIDOR   HEMLOCK FERN FOREST

10

09 James Corner/ Field Operations, *Waterfront Seattle*, 2014. Site plan and diagram of habitat areas and fish corridor.
10 *Waterfront Seattle*. Transect depicting the public promenade cantilevered over the salmon migration corridor, as well as upland habitat that includes prairie, pine forest, maple alder woods, and hemlock fern forest.

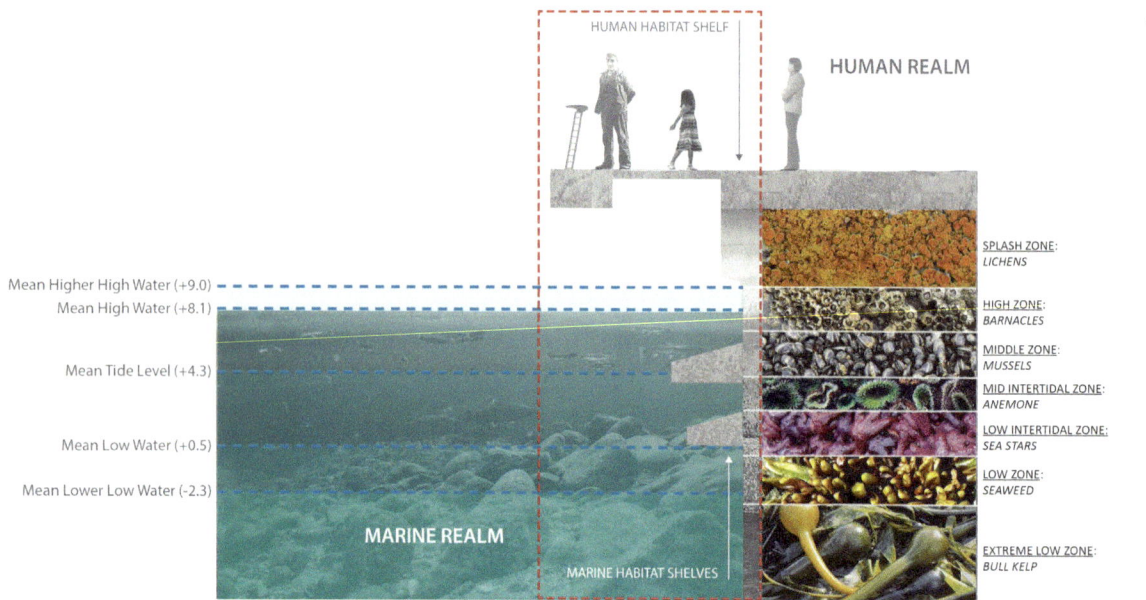

HUMAN HABITAT SHELF

HUMAN REALM

Mean Higher High Water (+9.0)
Mean High Water (+8.1)

Mean Tide Level (+4.3)

Mean Low Water (+0.5)

Mean Lower Low Water (-2.3)

MARINE REALM

MARINE HABITAT SHELVES

SPLASH ZONE:
LICHENS

HIGH ZONE:
BARNACLES

MIDDLE ZONE:
MUSSELS

MID INTERTIDAL ZONE:
ANEMONE

LOW INTERTIDAL ZONE:
SEA STARS

LOW ZONE:
SEAWEED

EXTREME LOW ZONE:
BULL KELP

11 Haddad|Drugan, *Seawall Strata*, 2015. Cross section at seawall depicting aquatic organisms that inform seawall pattern.
12 *Seawall Strata*. Models studying patterns for the seawall surface derived from abstractions of mussels and anemone.
13 *Seawall Strata*. Photograph at low tide of the seawall panel and habitat bench installed at Waterfront Seattle.

In *The Fish in the Forest*, Dale Stokes writes that salmon's "habitat-transforming efforts have evolved and shaped the coastal landscape over many thousands of square kilometers along the north Pacific rim."[27] Salmon are a keystone species in the Pacific Northwest of the United States and their importance to marine, riverine, and upland ecosystems cannot be understated. Their journey from riverbed to the sea, where they grow in marine environments, and their navigation back to the same river again to spawn and die, is a remarkable cycle. This cycle results in millions of pounds of salmon carcasses that provide an enormous quantity of nutrients to upland forests and their inhabitants.

Wild populations of Chinook, Coho, and Chum Salmon in this region were in decline in the twentieth century due to overfishing, destruction of habitat, and the construction of hydroelectric dams.[28] To bolster salmon populations, hundreds of hatchery facilities, operated by federal, state, tribal, and local governments were created. Unfortunately, the emphasis on artificial propagation of salmon has proven detrimental to naturally spawning salmon populations.[29] In 2005, a review of hatchery operations led to the recommendation that fish stocking be coordinated with salmon-habitat restoration, sustainable harvest, and hydroelectric-dam management to meet population goals.[30] The construction of a new salmon habitat at the urban waterfront in Elliott Bay, through the Waterfront Seattle project, transformed this recommendation into a reality.

The redesign of over 2,000 meters of the seawall into a biodiverse migration corridor necessitated a coalition of landscape architects, scientists, environmental consultants, engineers, and artists. Several features were proposed to increase the wall's function as habitat: intertidal habitat shelves to subdivide the face of the wall into shallow zones, textured surfaces to allow plants and invertebrates to cling, habitat benches that line the wall base, and an overhanging public promenade, constructed with glass block to allow light to penetrate the environment below.[31] The artist team Haddad|Drugan was commissioned to design the pattern of texture on the wall surface. They created stratified geometric textures that interpret the forms of the marine life—barnacles, mussels, anemone, starfish, and rockweed—intended to colonize the wall in rough alignment with where these lifeforms live in the intertidal zone.[32] As Haddad|Drugan explained, the seawall is "a dynamic seam where a myriad of elements, forces, and life forms interact in unique and complex ways."[33] The seawall became an enriched environment that crosses categories of use. It aligns a salmon migration corridor with a public promenade, and critical flood infrastructure functions as a substrate for public art.

Three kilometers away, the Olympic Sculpture Park by the architecture firm Weiss/Manfredi is another high-profile public landscape that enables salmon migration along the urban edge of Elliott Bay. The project includes a pocket beach and habitat benches that anchor a gradient of aquatic to upland environments. The Landscape Architecture Foundation reports the beneficial effect of this project on juvenile salmon populations, finding that "populations rose from just over 500 in 2007, to over 265,600 within the first three years of shoreline monitoring along the park's pocket beach and habitat bench."[34] Understood as part of a larger culture and history of salmon and waterway management in the Pacific Northwest, the Waterfront Seattle and Olympic Sculpture Park projects exemplify the emerging notion of urban waterfronts as multispecies and multifunctional landscapes. Notably, these strategies in Seattle bolster salmon populations and enable salmon agency at the same time, allowing these extraordinary fish to continue their work of shaping the landscapes and ecosystems of the region.

14 Thomas Chung, *Floating Fields*, 2016. View to one of dozens of aquatic basins designed for fish and vegetable polyculture and installed at the 2016 Shenzhen Hong Kong Urbanism/Architecture Bi-city Biennale. The project alludes to both the pervasive raft-based oyster culture in nearby Deep Bay as well as the historical dike-pond system of the Pearl River Delta.

Catalytic Polycultures

In the celebrated book *Four Fish: The Future of the Last Wild Food*, Paul Greenberg concludes his survey of stagnating yields in wild-catch fisheries and the rise of fish farming by promoting aquaculture that "starts from a place of polyculture . . . where systems instead of individual species are mastered."[35] For centuries, humans have in fact created systems like the polyculture Greenberg describes. The ancient polyculture of carp and rice in the terraces of East China exemplifies the construction of mutualistic relationships between fish and plant species. Fish feed on rice pests, and fish waste fertilizes rice growth. Rice plants, in turn, provide shade for carp. Rice-fish culture significantly reduces the quantity of fertilizers and pesticides needed to sustain rice monocultures (see Case Study 11). Ancient Hawaiians likewise developed mutualistic relationships between taro and freshwater fish in their farming practices. (see Case Study 15).

The benefits of polyculture are not just ecological. Polyculture practices showcase how designed mutualism can sponsor unexpected economies and change cultural behaviors. For example, the Seawater Energy and Agriculture System (SEAS) is a fish-plant polyculture project that joins aquaculture to the process of jet-fuel production. The project is located at Masdar City in Abu Dhabi, United Arab Emirates across from the runways of the Abu Dhabi International Airport and is managed by the Masdar Institute of Science and Technology. SEAS was designed as a replicable strategy for food and energy security in arid, coastal environments in this region and beyond. At SEAS, fish and shrimp are cultured in seawater ponds, and nutrient-rich effluents from aquaculture are used to irrigate fields of dwarf glasswort (*Salicornia bigelovii*), an oil-rich halophyte whose seeds are harvested to produce biofuel. The water then flows to a cultivated mangrove forest for water polishing and carbon sequestration before cycling back to the fishponds.[36] The first crop of dwarf glasswort was harvested in 2017 and the project is slated to expand.[37] Relative to the recent disciplinary interest in airport urbanism among landscape architects, SEAS opens up a surprising relationship between fish, cities, and flight.[38]

Another demonstration of polyculture as a far-reaching, transformative practice is Growing Power, the non-profit urban farm based in Milwaukee, Wisconsin. Growing Power has transformed vacant urban lots into productive aquaponic systems where 100,000 tilapia and perch are raised annually. At its peak in the early 2000's, the program supplied more than fresh food to underserved populations; it was an active agent in urban revitalization and brought attention to important issues of social justice. Growing Power hosted apprenticeship programs such as Youth Corps to educate and train a new generation of urban farmers in aquaponics, vermiculture, and making compost.[39]

The following projects build on the polyculture discussion presented by the examples of SEAS and Growing Power and explore the agency of polycultures that operate at multiple scales. Floating Fields, an exhibition landscape in Shenzhen, China, by Thomas Chung, adapted a derelict factory courtyard into a fish and vegetable polyculture, enlivened with public harvest and tasting events. A second project, Quabbin Fishery by Michael Ezban, brings polyculture to Boston's water supply infrastructure.

Floating Fields, by Thomas Chung

Floating Fields was a temporary productive landscape in Shenzhen, China, designed by Thomas Chung—Associate Professor at the School of Architecture at the Chinese University of Hong Kong—for the Shenzhen Hong Kong Urbanism/Architecture Bi-city Biennale of 2016 (UABB).[40] This landscape linked to the historical aquaculture ecologies of the Pearl River Delta (PRD) in Guangdong province, China, and illustrated how the creative deployment of polyculture might inform alternative food production and recreation at derelict properties and buildings in the rapidly urbanizing PRD.

For centuries, the PRD was home to an extensive, ecologically complex aquaculture landscape. The landscape was characterized by a mosaic of small fishponds, canals, and agricultural villages. Known today as the dike-pond system (DPS), this extraordinarily productive aquaculture-agriculture landscape featured a polyculture of fish, vegetables, fruit, and silk (see Case Study 10). At the end of the twentieth century, much of the DPS landscape was erased due to shifts toward intensive monocultures, the expansion of manufacturing industries, and rapid urban development in the region. More recently, as manufacturing has decamped from urban centers and as villages in peri-urban areas continue to evolve, the PRD has become a fragmented patchwork of high rises, urban farms, derelict industrial zones, and erased aquaculture.

Radical and ongoing economic, ecological, social, and formal changes to the contemporary landscape of the PRD informed the design of Floating Fields in Shenzhen. Floating Fields featured an interconnected series of rectilinear ground and roof ponds where fish, vegetables, and algae were grown. The project was sited at the service court and dormitory of the abandoned Da Cheng Flour Factory. Floating Fields utilized water from a buried waterway that was covered by the construction of the factory. Suppression of existing hydrologic conditions, such as the filling and covering of fishponds and canals to create a "neutral" substrate for construction, is a commonplace practice in the PRD.[41] Floating Fields tapped into the hydrological history of its site by aligning the array of its basins with the subsurface waterway. Visitors to the landscape could access and experience the water via steps, bridges, overlooks, and even small boats.

Although Floating Fields made reference to the historical DPS landscape, there are key differences between the ecological processes of the two aquaculture-agriculture systems. The DPS was characterized by independent fishponds structured by earthen dikes. The dikes were planted with crops and fertilized with pond mud that was enriched with fish waste. Inherent to the functioning of the system was the excavation and placement of mud. Floating Fields, on the other hand, was a soil-less hydroponic system that relied on a flow-based system of recirculating water for nutrient transfer. Nutrient-rich "waste" water, from carp and duck farming, fertilized crops planted on rafts that floated across the surface of the ponds. Water would then flow into the algae production ponds to fuel the growth of food for the fish. After the water was cleansed in the filtering ponds, and purified in the water lily pond, it was directed back to the fishponds. Despite differences in their form and ecological processes, the polyculture at both the DPS and Floating Fields was spatially organized by biological interactions between fish and plants.

The term "designer ecology," described by ecologist Nina-Marie Lister, is a helpful lens through which to appreciate Floating Fields. Designer ecologies are "vital, indeed essential, for educational, aesthetic, spiritual, and other reasons. . . . Designer ecology, while valid and desirable in urban contexts for many reasons, is not operational

15

1	Water Lily Pond
2	Carp Pond
3	Duck Pond
4	Floating Fields • Aquaponics
5	Floating Fields • Mulberry Fishpond
6	Silkworm Pavilion
7	Mulberry
8	Vegetables
9	Algae Pavilion & Ponds
10	Filtering Ponds
11	Rooftop Cultivation
12	Wheat
13	Workshop

16

17

15 *Floating Fields*. Axonometric diagram of ground level and rooftop basin arrangement and function.
16 *Floating Fields*. View to aquatic basins, terrestrial plots, and floating planted rafts.
17 *Floating Fields*. Children participating in the capture of carp at the Harvest Festival during the final days of the Biennale.

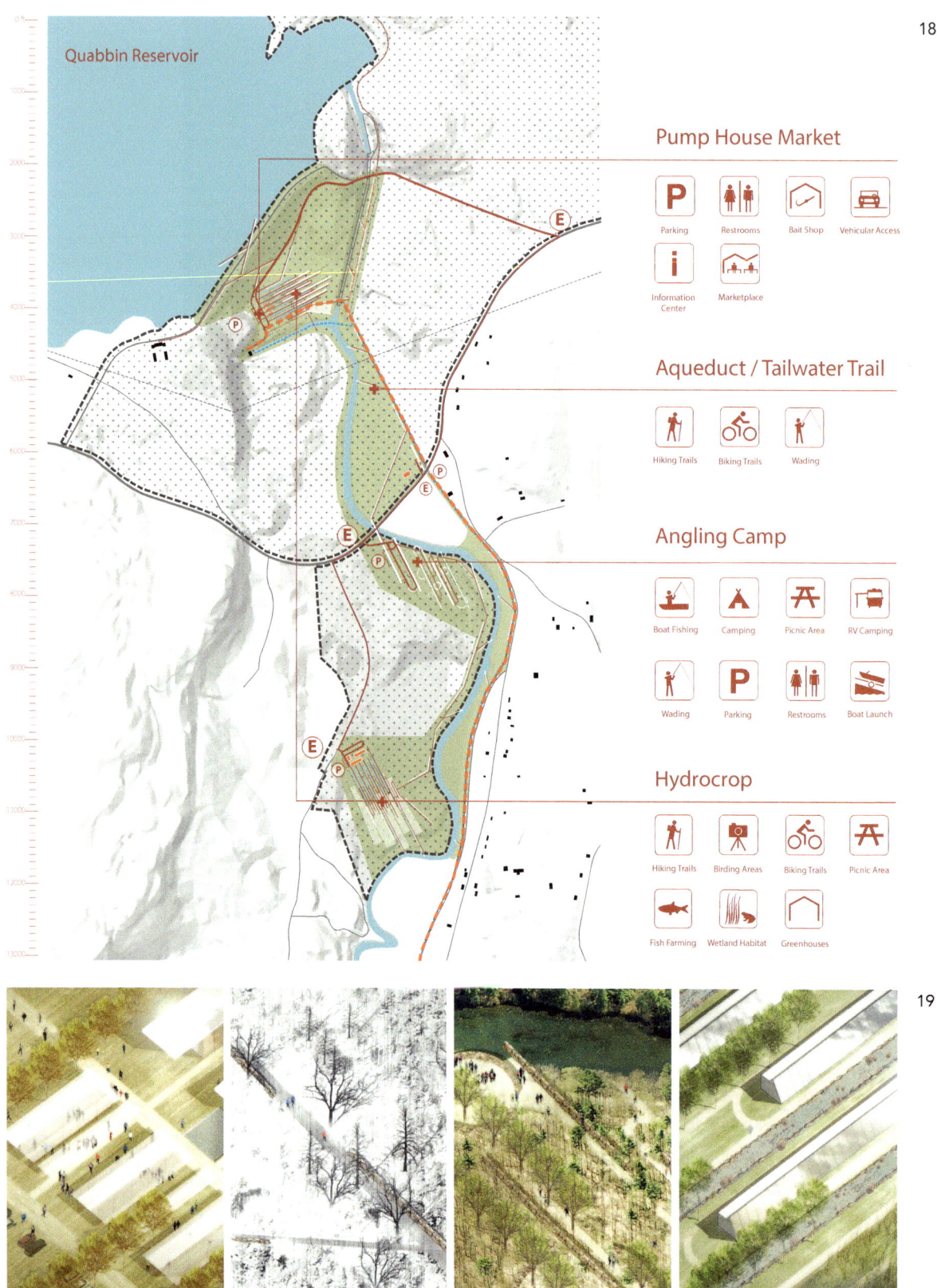

Pump House Market

Parking	Restrooms	Bait Shop	Vehicular Access
Information Center	Marketplace		

Aqueduct / Tailwater Trail

Hiking Trails	Biking Trails	Wading

Angling Camp

Boat Fishing	Camping	Picnic Area	RV Camping
Wading	Parking	Restrooms	Boat Launch

Hydrocrop

Hiking Trails	Birding Areas	Biking Trails	Picnic Area
Fish Farming	Wetland Habitat	Greenhouses	

Quabbin Reservoir

18 Michael Ezban, *Quabbin Fishery*, 2014. Site plan depicting four zones of the Quabbin Fishery, a proposed 3 km long, 100 ha productive tailwater fishery. The fishery occupies portions of public land directly adjacent to the Swift River.
19 *Quabbin Fishery*. Renderings of the Pump House Market, Aqueduct/Tailwater Trail, Angling Camp, and Hydrocrop.

ecology; it does not program, facilitate, or ultimately permit the emergence and evolution of self-organizing, resilient ecological systems."[42] Floating Fields was a highly managed, closed-loop designer ecology that was specific to its site in Shenzhen, yet it was expansive in that it turned the ecological principles of sustainable fish farming into an imaginative public experience with historical resonance. During the months of the UABB, planting, tasting, and harvest festivals offered participants opportunities for sensorial engagement with this polyculture.

In her article, "Urban Agriculture in the Pearl River Delta," urban theorist Margaret Crawford examines the social, political, and economic forces behind the transition of agricultural villages into high-density urbanism in the PRD. Crawford envisions alternative food production futures for these transitional zones and she points to "the rapidity of change and the appearance of new ideas and initiatives" around urban agriculture in this region.[43] Additionally, she writes that landscape architects and designers have particular agency in urban agriculture; they add value to agricultural pursuits by "giving new cultural and aesthetic meanings to the practicalities of growing food."[44]

Floating Fields was an exemplary effort that furthers the work of designers, planners, and theorists who envision food and farming in the transitional spaces of urbanization across the PRD. Projects like this one highlight the historical significance of constructed productive ecologies in the region and model how urban polycultures can offer much needed pushback against what have become the normative processes of urbanization in the PRD in the twenty-first century.

Quabbin Fishery, by Michael Ezban

Quabbin Fishery is a design proposal by the author that was developed with support from the Maeder-York Family Fellowship in Landscape at the Isabella Stewart Gardner Museum, in Boston. The proposal reimagines fish culture at the Quabbin Reservoir, a massive lake constructed in the 1930s in central Massachusetts, United States, to supply Boston with drinking water. Quabbin Fishery is designed to demonstrate scalable and flexible strategies for polyculture and aquatic habitats that would be applicable to other reservoirs, tailwaters, and aqueducts in Boston's vast water-infrastructure network.

Quabbin Reservoir was created by damming the Swift River. Damming rivers, a ubiquitous twentieth-century practice in the United States for water supply and energy creation, also happens to create optimized trout habitats. Once a dam is built to stop up the flow of a river, a significant ecological shift occurs. The river that exits the reservoir changes from a freestone river to one that draws water from the depths of the reservoir. Trout thrive in these so-called *tailwaters* that run at cold temperatures in the summer and warmer-than-usual temperatures in the winter, and are regulated to flow at consistent volumes throughout the year. Due to vegetal decay and seasonal overturning at the reservoir, tailwaters become rich with nutrients that support the growth of aquatic weeds and insects on which trout feed.[45]

In the United States, most trout that inhabit tailwater ecologies originate from the massive infrastructural system of fish hatcheries. Today, the Fisheries Program of the United States Fish and Wildlife Service, established in 1871, operates a network of seventy National Fish Hatcheries that annually produce and distribute millions of fish and eggs across the country, primarily for recreational purposes (see Part Three). This decentralized system cooperates with hundreds of state and tribal hatcheries. Starting in the nineteenth

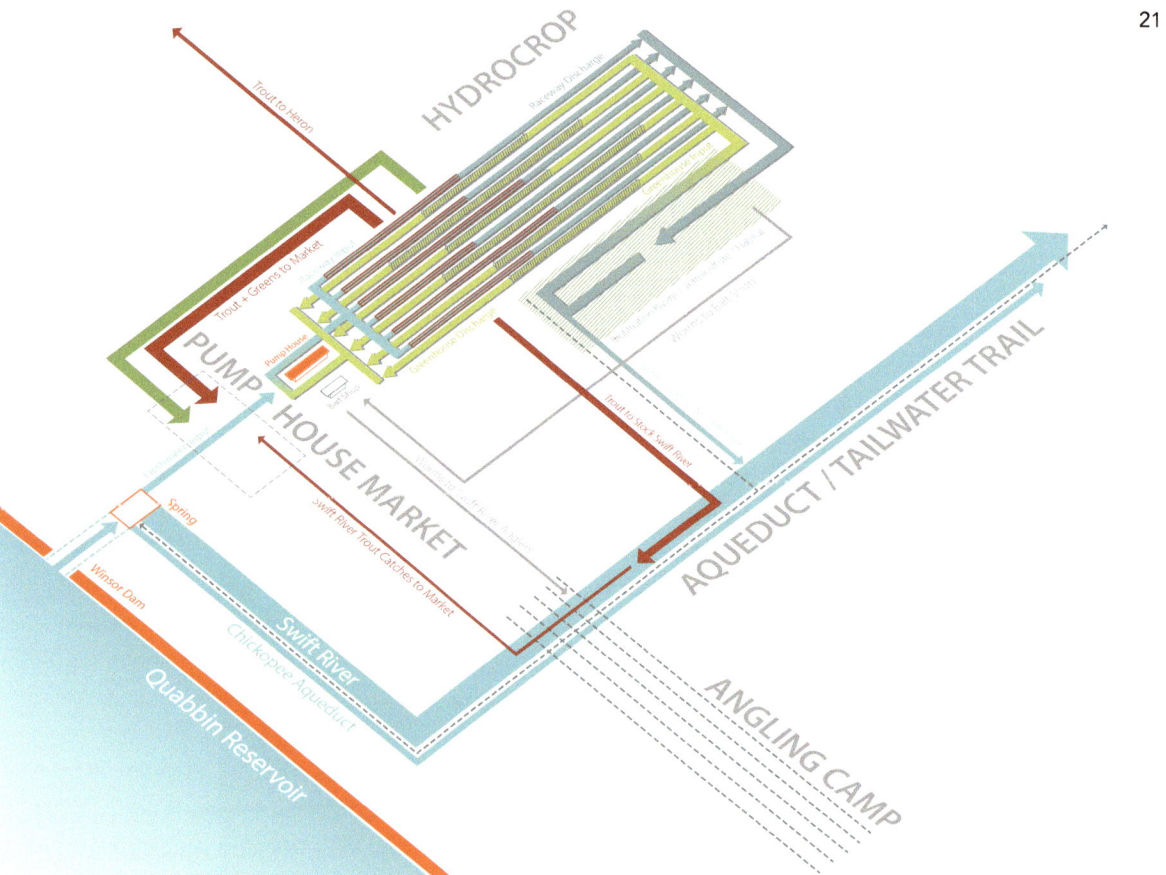

20 *Quabbin Fishery*. Section at Aqueduct/Tailwater Trail depicts stone spurs and walls that structure the riverine landscape.
21 *Quabbin Fishery*. Diagram of the hydrologic flows and material exchanges that link rivulet, wetland, tailwater, aqueduct, and reservoir at Quabbin Fishery.

century, the Rainbow Trout (*Oncorhynchus mykiss*), a fish that is native to the Pacific Rim and prized among anglers for its fight, coloration, and taste, has been the premier stocked fish nationwide. In *An Entirely Synthetic Fish*, which is a history of Rainbow Trout, Anders Halverson claims that approximately "100 million of them leave the hatcheries every year weighing about a quarter pound apiece—a total of 25 million pounds of Rainbow Trout dumped into America's freshwaters."[46] Tailwaters such as the Swift River, which is flowing with cold, nutrient-rich water and coursing with stocked trout, are constructed sites of encounter for millions of anglers and farmed fish.

Quabbin Fishery operates within the socio-ecological system of the Swift River and the adjacent McLaughlin State Fish Hatchery that routinely stocks the tailwater with trout. The hatchery, and its trout monoculture, is recast in this project as a biodiverse, aquaponic landscape that joins trout cultivation, greenhouse agriculture, and a wetland habitat. The Quabbin Fishery design also proposes an adaptation of the Swift River as a new spawning habitat intended to foster self-sustaining trout populations. These populations would gradually reduce the need for stocking in the future. The newly constructed trout stream at Wolf Creek National Hatchery in Kentucky, a pioneering effort to support self-sustaining trout populations alongside traditional production and stocking, was the model for the design of these Swift River transformations (see Case Study 14).

The Quabbin Fishery plan occupies 100 ha of public lands that abut the Swift River, extending from Winsor Dam at Quabbin Reservoir to the McLaughlin State Fish Hatchery, that is three kilometers downstream. The design includes four intermittent riparian landscapes where aquaculture and angling are set among mixed hardwoods, conifers, and meadowlands. Two of these landscapes, the Hydrocrop and the Aqueduct/Tailwater Trail, exemplify the new types of habitats proposed for the Swift River.

Hydrocrop converts the McLaughlin Hatchery into a mixed aquaculture, agriculture, and recreational space. This landscape is characterized by four interdigitated elements: 1) rows of bare-concrete trout raceways that have been remade into rivulets laid with beds of rock and gravel; 2) a parallel pattern of linear greenhouses; 3) intermittent lines of sugar maple trees (*Acer saccharum*), that shade winding trails and bike paths; and 4) terraced basins for wastewater filtration that also provide habitat for avian species.

In Massachusetts, trends since the early 2000's indicate growing economies of fish and greenhouse-vegetable production, agritourism, angling, and bird watching.[47] The increasing economic value of these programs and land uses in Massachusetts was one impetus to explore aquaponic landscapes as an additional program for increasing state revenues in the Quabbin Fishery proposal. The proposal assumes an approach that follows the model of public-private partnership, such as at the Veta la Palma fish farm in Spain, where public land is leased for sustainable commercial aquaculture (see Case Study 03).

The Aqueduct/Tailwater Trail is a proposal for a recreational corridor that runs alongside the tailwater. The marked trail facilitates access to constructed off-channel streams and gravel riffles that would provide habitat and spawning zones for trout. Stone spurs constructed in the tailwater to reduce bank erosion extend far up the bank and create a material connection between the terrestrial trail and the aqueous world of trout.

Anne Whiston Spirn writes, "There is always a tension in landscape . . . between the human impulse to wonder at the wild and the compulsion to use, manage, and control."[48] The constructed ecologies, assembled species, and fabricated habitat of Quabbin Fishery fall somewhere on the gradient between wild and managed.

22 hutterreimann + cejka Landschaftsarchitektur, *Landesgartenschau Wernigerode 2006*, 2006. The Fish Walk, a continuous 1 km long boardwalk that spans seven historical fishponds and a landfill in Wernigerode, Germany. A variety of follies, such as the Folly WhiteBox seen here, are intermittent features found along the Fish Walk.

Post-Aquaculture Adaptations

Landscape theorist Sebastien Marot invokes the term "anamnesis," or recollection of previous history, to describe a process of site analysis and landscape design that is attuned to what came before. For Marot, site analysis informed by anamnesis starts with a reading of the site "as a palimpsest that evidences all of the activities that contributed to the shaping of that particular landscape."[49] In the context of landscape design, a palimpsest is often considered as the material registration of temporal and spatial events in a landscape.[50]

Like the landscapes about which Marot writes, aquaculture practices have lasting material effects on their sites. Former fish farms evince the specific processes of farming and cohabitation that at some sites extend back centuries. Residual hydrologies, topographies, and ecologies of former farms have great potential to shape future uses and readings of the site. For instance, at the tidal shrimp farms of the Mai Po Marshes in Hong Kong, the manipulation of water levels that once facilitated aquaculture continues, even though commercial shrimp production is no longer the priority. The hydrological practice is maintained in service of a new goal—to optimize habitats and create feeding opportunities for a variety of migratory and endangered bird species (see Case Study 12). The brackish basins constructed centuries ago for eel harvests at Comacchio, Italy, serve in the twenty-first century as a biodiverse coastal lagoon within the park system of the Po River Delta (see Case Study 04).

Landscape architect James Corner, framing landscape through the lens of hermeneutics, also speculates on latency in landscape. Like Marot, he describes landscape as a "topographic palimpsest" of "collaged and weathered overlays."[51] Corner also writes on the agency of a site's history in transforming its future. "Residua," he observes, "provide loci for remembrance, renewal, and transfiguration of a culture's relationship to the land."[52] Landscape architects that adapt fish farms for new uses, and new populaces, make choices about the degree to which to preserve or alter the material legacies of aquaculture on the site. Through this work, landscape architects become part of an ongoing process of change that will inform future interpretations and anamnesis of aquaculture on the site.

The two projects discussed next delve further into the specific ways that landscape architects adapt the residua of former fish farms to create new, public aquaculture landscapes that are enriched with history. The first project, *Landesgartenschau Wernigerode 2006*, by hutterreimann + cejka Landschaftsarchitektur, interweaves seven historical fish ponds for a garden show and public landscape in the town of Wernigerode, Germany. A second project, Yichang Yunhe Park, by Turenscape, transforms a derelict fish farm into a stormwater management landscape for the city of Yichang, China.

Landesgartenshau Wernigerode 2006, by hutterreimann + cejka Landschaftsarchitektur

Hutterreimann + cejka Landschaftsarchitektur (hr+c) designed a public landscape that interprets and uncovers histories of aquaculture in the town of Wernigerode, in the Harz region of Germany. The project features a remarkable promenade and various follies sited at a series of historical fishponds. While the ponds no longer support commercial aquaculture, they are relics of a long history of pond-based carp farming in central and eastern Europe that dates back to medieval times.[53]

In 2002, hr+c won the design competition established by the town of Wernigerode for a garden exhibition named *Landesgartenschau Wernigerode 2006*. Competition participants were charged with the task of designing a landscape that would serve the garden exhibition as well as function as a recreational space for the town after the show was complete. The site selected by the town for the exhibition was in a transitional area and was characterized by prefabricated flats, a rubble landfill, and post-industrial land. The most notable feature of the site was the chain of seven derelict, historical fish ponds that were encircled by swaths of spontaneous vegetation. Hr+c's design objective was to revitalize the ponds through a reading of this neglected and peripheral zone of Wernigerode as an "overlay of natural and artificial traces."[54]

The seven existing fishponds at Wernigerode range in size from 0.25 ha to 5 ha. The first of these ponds is thought to have been constructed in the fourteenth century. The ponds are arranged in a northeast/southwest orientation, and most are separated by earthen dikes that are approximately 6 m in width. The largest swath of land between ponds is the site of a roughly three-hectare landfill that is filled with construction rubble and has served as a dumping ground since the mid-twentieth century.

Hr+c introduced a Fish Walk promenade in their design that is intended to enable embodied, sensorial experiences of the ponds and their histories—the Fish Walk "leads to the stories of the fishponds."[55] The Fish Walk is a 1 km long raised walkway that forms a continuous backbone of the landscape by linking all seven ponds together and threading through a cleft in the landfill. This promenade weaves together the historical site programs, the contemporary ecologies and pond environments, and the elements of the garden exhibition. The Fish Walk is adjoined by a series of "follies" that were designed in collaboration with the architecture firm, A_lab architektur.

The follies provide opportunities for full-sensory engagement with the water, close observation of the fish habitat, and surreal displays of fish communities. The Folly Fischzucht, or fish culture, displaces and displays samples of the benthic environment of the ponds, as well as some of its fish, in illuminated glass boxes that are arranged on a wooden sideboard at the edge of Köhlerteich pond. Further along the Fish Walk, visitors encounter built-in telescopes for close observation of animal life within the ponds' edge ecologies. The most curious folly along the Fish Walk is the Folly WhiteBox—an immersive exhibition in a darkened room that hovers over Kurtsteich pond. In the WhiteBox a projected, looping film of underwater footage of fish life encircles visitors, creating a surreal environment that offers an ichthyocentric perspective into farmed fish habitat.

The Fish Walk and follies are not the only design elements that allude to and interpret histories of fish farming in Wernigerode. The use of limestone by hr+c in their landscape design also creates a poetic reference to historical aquaculture practices. Though not noted as such by the landscape architects themselves, the limestone resonates with the ubiquitous aquaculture practice of using limestone to regulate the pH

23

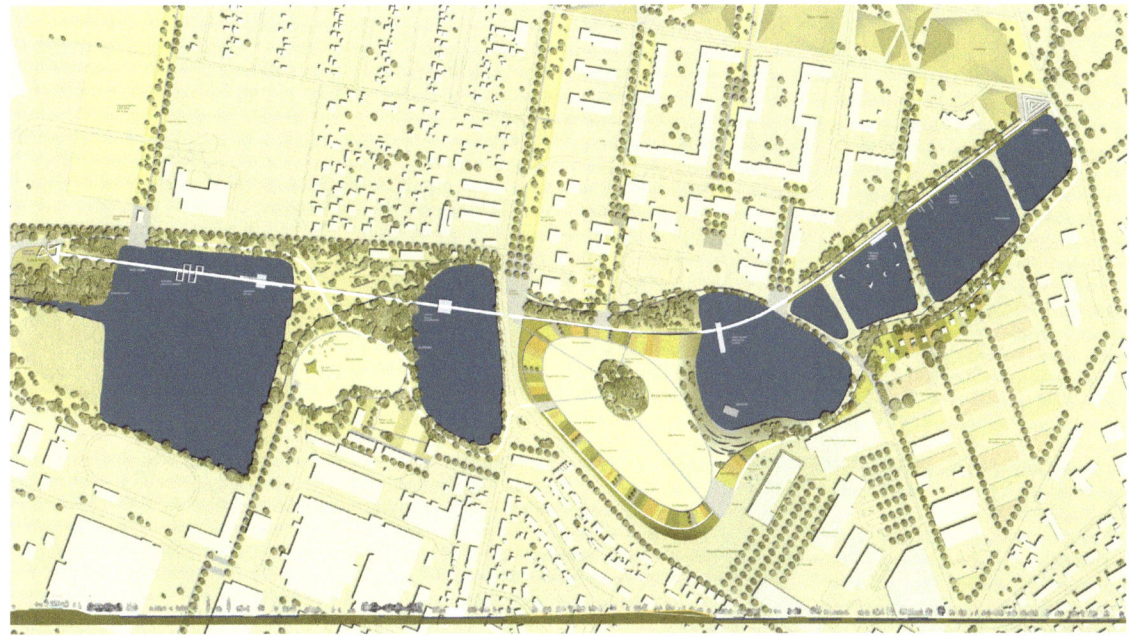

24
25

23 *Landesgartenschau Wernigerode 2006.* Site plan depicts Fish Walk across seven fishponds and an array of garden plots.
24 A_lab architektur, *Folly Fischzucht, 2006.* An aquarium at edge of Kohler fishpond displays a variety of fish species.
25 A_lab architektur, *Folly WhiteBox, 2006.* A virtual underwater experience that offers visitors an ichthyocentric perspective.

26 Turenscape, *Yichang Yunhe Park*, 2014. Satellite photographs of derelict fish farm (left) and completed public park (right). The reshaped ponds, planted trees, and circulation infrastructure are evident in the image of the contemporary park.
27 *Yichang Yunhe Park*. Stairs and terraces provide visitor access to the stormwater held in the former fishponds.

of the pond water. Regularly casting crushed limestone into the water softens the acidity and also prevents the pH from changing rapidly. The practice is common in countries across the world, and was likely employed by the fish farmers of these ponds.

Hr+c designed an evocative display of limestone at the intersection of the Fish Walk and a cleft in the landfill that they call a "mineral gorge." Lining the canted sidewalls of the gorge are gabions with meticulous arrangements of Harz limestone collected from the geologically diverse region of Wernigerode. The mineral gorge is not the only instance where limestone is featured in the landscape design. Paths are also surfaced with crushed limestone, putting visitors in touch with this material with each step. Through the creative deployment of limestone in their landscape design, hr+c encourages visitors to trace a specific material thread that connects ancient geologic processes, historical aquaculture practices, and the program of the contemporary exhibitions.

Landscape historian John Dixon Hunt suggests that, "The designer must surely be, in some form, a historian in getting to understand the given site, so that what s/he does, as an implied historian, is to make palpable whatever history s/he has discovered there."[56] Hr+c make their ambition to act as historian designers clear in their goals to awaken "dormant potential, revealing hidden qualities without erasing the traces left of history."[57] The Fish Walk reveals and interprets more so than it reinvents the centuries-old fishponds of Wernigerode.

Yichang Yunhe Park, by Turenscape

The Yichang Yunhe Park, an urban park in Yichang, Hubei province, China, is a remarkable example of a fish farm adapted into public stormwater infrastructure. The project, designed by the Chinese landscape firm Turenscape and opened to the public in 2014, exemplifies the firm's aspirations to enrich the cultural value and the performative functions of public landscapes in China with references China's agricultural heritage.[58] It provides for the infrastructural needs of the city by capturing and filtering stormwater prior to releasing it into the Yangtze River. The park also provides recreational experiences and ecological services for the city. Turenscape identified the infrastructural potential of this landscape as part of a commission to produce broad ecological planning for the city.

Prior to its transformation into a public landscape, the site was a 12 ha commercial fish farm. Denuded earthen dikes subdivided the kidney-bean shaped farm into twelve fishponds of varying size. The farm was bordered to the south by a busy urban street and to the north by an arching urban canal that dates to the mid-twentieth century. The farm discharged fishpond effluent, which was high in nutrients that contribute to poor water quality, into the canal, and this water ultimately reached the nearby Yangtze River. Today, housing towers, built in tandem with the construction of Yichang Yunhe Park, loom over the landscape.

Turenscape's approach to adapting the fish farm to manage and filter urban stormwater includes surgical topographic modification and targeted planting strategies. Strategic openings are created in the dikes to interconnect the ponds and allow for the stormwater to pass from one pond to the next. Contaminated water from the canal enters at the easternmost pond and flows through the other ponds before discharging back into the canal at the other end of the park. The ponds slow and hold stormwater so that solids can settle. The water also passes through wet ponds that are planted with lotus, cattails, and reeds that absorb excess nutrients and chemical pollutants.

Alongside the infrastructural transformations, Turenscape also developed an imaginative circulation and tree planting strategy. The dikes that shape the ponds were thought of by Turenscape as armatures for public circulation that edge the fishponds. Turenscape also offers ways to depart from the historical landscape pattern. Visitors can descend from the tops of the dikes via stairways that connect to observation platforms at the water surface. Groves of trees, including the dawn redwood (*Metasequoia glyptostroboides*) that is native to nearby Hubei province, were planted at the dikes and water edges to create shaded areas where visitors can rest and reflect. Visitors may also ascend to a 410 m long "skywalk." The skywalk is a whimsical, sinuous ribbon that threads through tree canopies and sits in contrast to the web of linear paths on the dikes below.

In 2015, the Sponge City Initiative was launched by the Chinese government. It is a national program that seeks to reduce the intensity of rainwater runoff by enhancing and distributing absorption capacities more evenly across targeted areas.[59] What makes Turenscape's contribution to the water infrastructure initiatives in China unique is that they imbue their designs for performative stormwater infrastructure with historical and cultural resonance. Adapting and recreating the patterns and processes of agricultural landscapes in their schemes results in landscapes that also have an aesthetic performance.[60]

In China, aquaculture, agriculture, and regional resiliency were historically intertwined. Fishponds, paddies, bunded terraces, polder landscapes, and irrigation canals have all been part of the extensive hydrological infrastructure that has structured and irrigated rural landscapes in China for millennia. Turenscape's principal, Kongjian Yu, reflects on these traditional landscapes: "For thousands of years, farmers managed living landscapes using the survival skills passed on by their ancestors through countless trials and errors. Generations adapted to both the threat and the results of natural disasters—floods, droughts, earthquakes, landslides, and soil erosion—while honing their abilities in field grading, irrigation, and food production."[61] In the twenty-first century, Turenscape brings productive landscape patterns and practices to their commissions for urban landscape design to achieve "a new aesthetic, grounded in appreciation of the beauty of productive, ecology-supporting, survival-enhancing things."[62]

The Yangtze River basin in particular is characterized by a deep history of agriculture that provides regional-scale infrastructural services alongside food production. During China's legendary first dynasty, for example, extensive systems of dikes and canals channeled Yangtze River overflow toward rice paddies for slow absorption and flood management.[63] In the twenty-first century, the Yangtze River basin is the hub of China's freshwater fish production—this region supplies more than half of the farmed fish in China. But aquaculture here is a polluting practice that threatens the river's ecologies.[64]

Turenscape's adaptation of a fish farm into a public landscape that remediates contaminated water extends the millennia-old history of infrastructural farms along the Yangtze River. As landscape theorist Kelly Shannon writes, Turenscape's integration of agriculture, infrastructure, and aesthetics is "evolutionary," and "the logical next step in a long tradition of Chinese land and water management."[65] Of the many public landscapes that Turenscape has designed, the adaptation of a fish farm in China—the global leader in fish production—is of particular importance. Yichang Yunhe Park models an alternative to polluting aquaculture practices,[66] and it illustrates that leveraging the rich histories and material legacies of fish farming in China can produce public aquaculture landscapes that enable aesthetic experiences, environmental remediation, and resilient urbanism.

28 *Yichang Yunhe Park.* View to path among dawn redwood trees, an endangered species native to nearby Lichuan county.
29 *Yichang Yunhe Park.* Stormwater enters the park at the easternmost basin (at the left in this photograph) and discharges into the canal at the westernmost basin (at the right). Terraces parallel to the road south of the park intercept surface runoff.

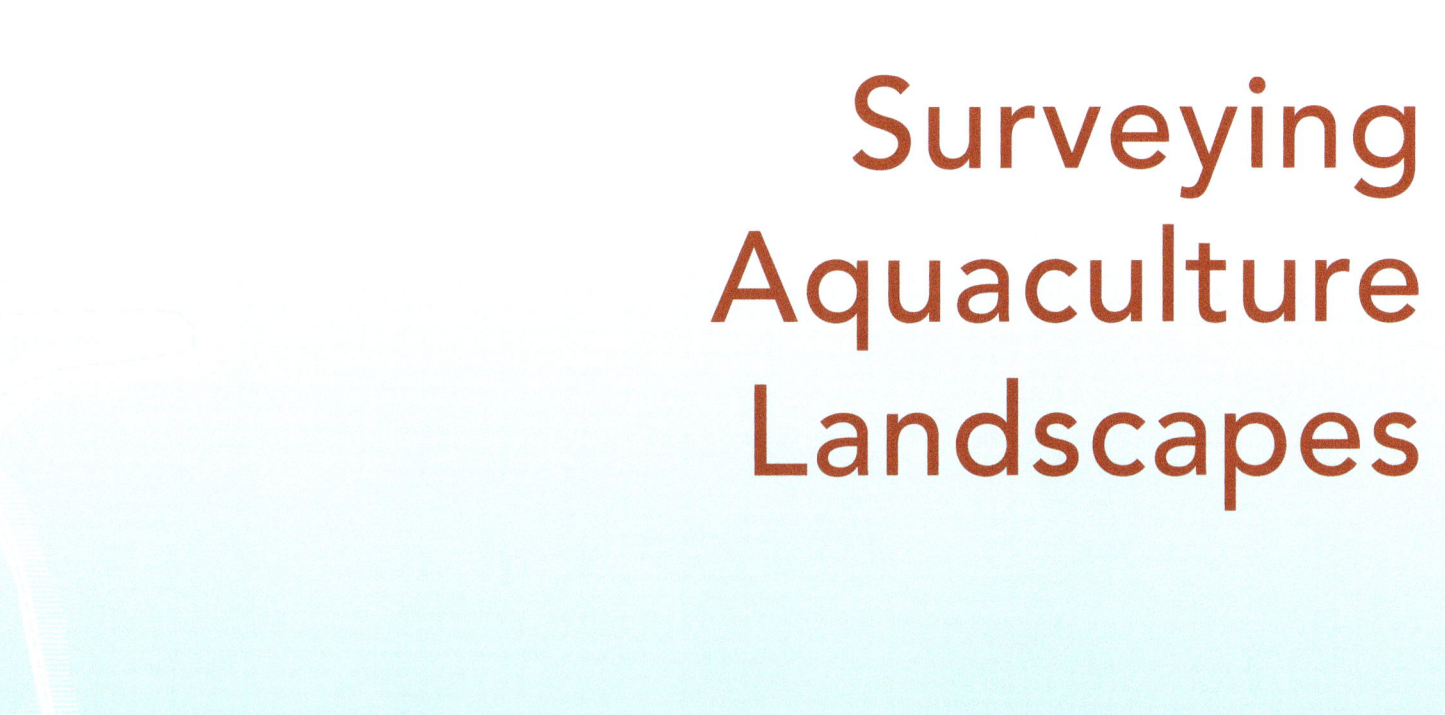

PART TWO

Surveying Aquaculture Landscapes

Case Study Geographic Distribution

Hawaiian Islands, USA

Wolf Creek Dam, USA

Baures, Bolivia

Doñana Nature Reserve, Spain
Bay of Cádiz, Spain
Paris, France
Lake Nokoué, Benin
Huningue, France
Comacchio, Italy
Sperlonga, Italy
Třeboň Basin, Czech Republic

Kolkata, India

Pearl River Delta, China
Mai Po Marshes, Hong Kong
Longxian Village, China

Case Study Timeline

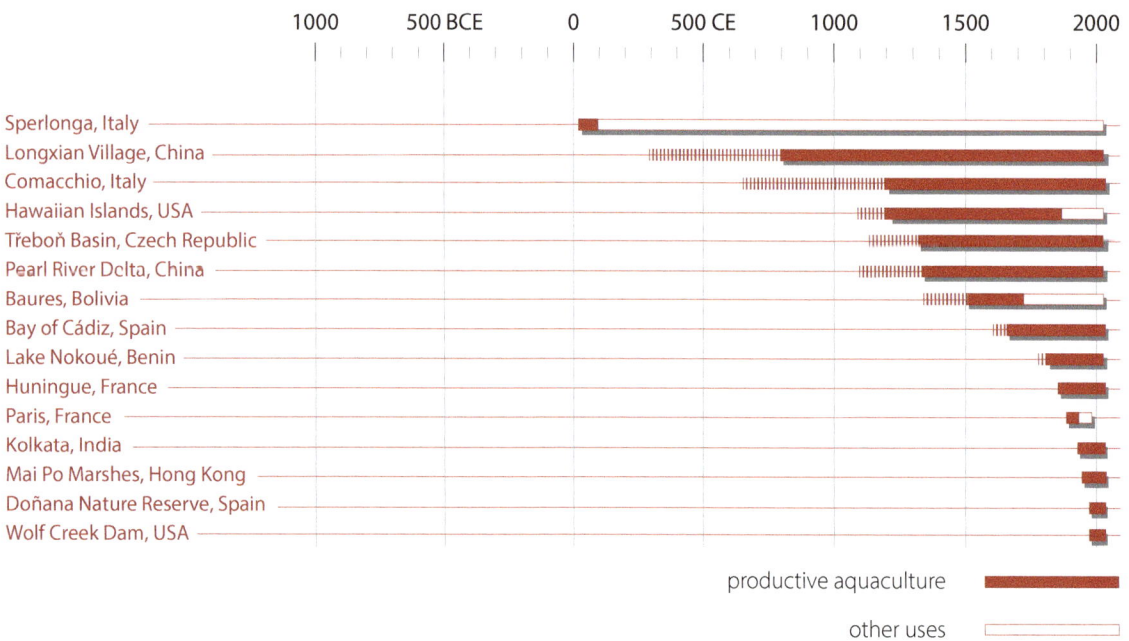

1000   500 BCE   0   500 CE   1000   1500   2000

Sperlonga, Italy
Longxian Village, China
Comacchio, Italy
Hawaiian Islands, USA
Třeboň Basin, Czech Republic
Pearl River Delta, China
Baures, Bolivia
Bay of Cádiz, Spain
Lake Nokoué, Benin
Huningue, France
Paris, France
Kolkata, India
Mai Po Marshes, Hong Kong
Doñana Nature Reserve, Spain
Wolf Creek Dam, USA

productive aquaculture

other uses

01 Case study landscapes are geographically distributed across five continents and ten countries. The case study timeline indicates the dates during which aquaculture was the primary function. The dates in which aquaculture practices commence at several of these landscapes is uncertain, and a dashed line is used to indicate possible start dates.

Overview of *Aquaculture Landscapes* Case Studies

The fifteen case studies collected in Part Two of this book describe a broad range of vernacular aquaculture landscapes. These landscapes are located across five continents and ten countries, and the majority fall within the tropical and temperate zones of the Northern Hemisphere. The case studies span many centuries, with some landscapes remaining active sites for aquaculture for much of that time. Others were farmed and abandoned long ago, only to host new land uses in the twenty-first century. Still other landscapes were only recently constructed for aquaculture as well as a variety of other functions. The sizes of these case study landscapes vary from less than one hectare to over 56,000 ha. They are situated in a range of contexts across urban-rural gradients from agricultural zones to peri-urban conditions to the hearts of cities. They are located within coastal lagoons, at estuaries, in upland savannas and forests, and along mountainsides.

Three factors informed the selection of this diverse group from the many available case studies. The primary factor is to investigate fish farms that are actual landscapes rather than indoor facilities or cage-based farms in marine environments. A second area of focus concerns multifunctionality. The landscapes selected for this study all support more than just the production of fish. Additional functions include ecological, infrastructural, and recreational services, which are either concurrent with aquaculture practices or emerge as post-aquaculture functions. The third factor is to explore fish farms that shape the public realm and are engaged by numerous constituencies. These public landscapes invite aesthetic experiences with fish through participatory conservation and harvest events, they feature recreational infrastructure and are integrated into public parks, and some are constructed on state-owned land and structured by public-private partnerships. The largest of these landscapes shape the public realm by virtue of their sheer size; they organize the settlement patterns of towns, cities, and regions.

Case Study Format

The case studies in this book are organized with consistent modes of description, analysis, and representation. The uniformity of graphics and writing facilitates comparisons across a broad range of fish farms. Each case study includes the following elements:

Key Data—Each case study is quantified by a landscape area, type of aquaculture practiced, aquaculture yield, and characteristics of water utilized for farming. To accommodate the fact that the size and productivity of some long-lived landscapes have varied considerably over time, metrics from different eras are listed where appropriate.

Description—A brief written description is provided to characterize each landscape type, history, and ecological and cultural context(s). Formal characteristics of the landscape are highlighted alongside its multifunctional traits, its modes of public integration, and its key organizational strategies and processes.

Images—Each case study features contemporary and historical images of the landscape, including photographs (ground level, aerial, satellite) and etchings.

Drawings—Featured in the case studies are four types of original drawings by the author that examine context, form and pattern, character, and performance of the landscape at a variety of scales:

Regional Diagram—A planimetric diagram ranging in scale from 1:300 m² to 1:300 km² that illustrates regional context.

Site Diagram—A planimetric diagram ranging in scale from 1:90 m² to 1:3 km² that illustrates landscape form, pattern, and immediate adjacencies.

Transect—A rendered axonometric projection of a 40 m x 80 m swath of the landscape that illustrates topographic and planted forms, materiality, texture, and color. The drawing also identifies key flora and fauna species.

Landscape Systems and Strategies—A range of diagram types at multiple scales illustrates the topographic, hydrologic, and vegetal strategies at work in each landscape. They include the distribution of key system elements; hydrologic and material flows, exchanges, and cycles; evolution of landscape form and function; seasonal practices and processes; and mutualistic relationships between fish, plants, humans, and other animals. Various modes of public integration are also described, including recreational itineraries and diverse programming.

Design Strategies in the *Aquaculture Landscapes* Case Studies

Each of the aquaculture landscapes described here is connected to circumstances, cultures, and histories. Their design is informed by climate and geologic conditions as well as the ecosystems in which they are embedded. They are socially constructed landscapes that are also physically coshaped by the fish and humans that inhabit them.

Despite their specificity, these landscapes offer design strategies that have applications beyond their respective contexts. The most pervasive theme in these case studies is the idea that the fish farm is managed as a habitat, where fish are free to express species-specific behaviors within biodiverse environments and human inputs are relatively limited. This practice, more commonly known as "extensive aquaculture," can be found in the projects covered in Case Studies 01, 02, 03, 04, 08, 09, 12, 13, 14, and 15.

Other design strategies are also present in the case studies. The colocation of different types of production practices diversifies economies at the salt and fish farms in the Bay of Cádiz in Spain (see Case Study 02). Animal-plant mutualism creates productive fish-rice polycultures in Zhejiang Province of China (see Case Study 11). Adaptive reuse is common, as demonstrated by a defunct shrimp farm in the Mai Po Marshes in Hong Kong converted into an ecological reserve (see Case Study 12). Wastewater treatment and aquaculture are integrated at the municipal and the household scale at sites in India and China (see Case Studies 09 and 10). Public-private partnerships in which states lease public land for environmentally regulated commercial fish farming can be found in the Třeboň Basin of the Czech Republic and Doñana Nature Reserve in Spain (see Case Studies 01 and 03).

The collection of remarkable landscapes illustrated here—which includes the atmospheric grotto aquarium in Paris, France, the ingenious fish traps that refigured the Hawaiian Island coastlines, the centuries-old savanna weirs in Bolivia, and many more—is a call to visionary aquaculture designers in the twenty-first century to look to precursors for vernacular wisdom and inspired strategies. These case studies offer a transcultural history of ideas to underlie the contemporary project of farming fish.

Case Study Area Comparison

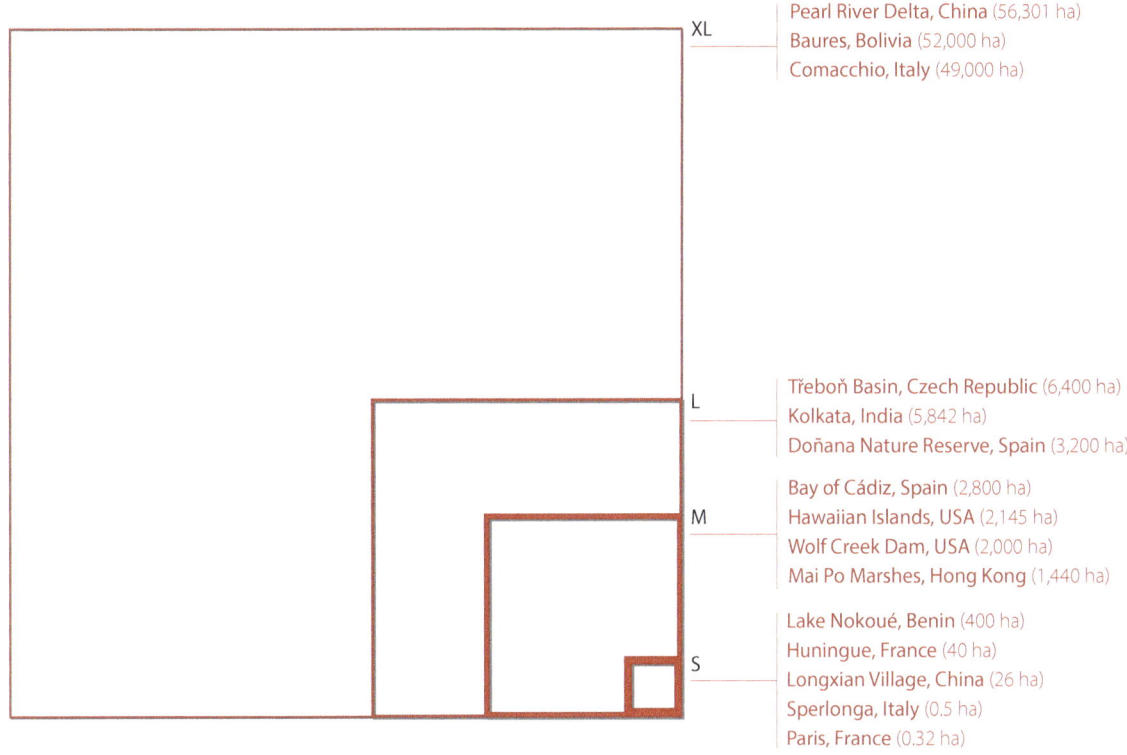

XL
Pearl River Delta, China (56,301 ha)
Baures, Bolivia (52,000 ha)
Comacchio, Italy (49,000 ha)

L
Třeboň Basin, Czech Republic (6,400 ha)
Kolkata, India (5,842 ha)
Doñana Nature Reserve, Spain (3,200 ha)

M
Bay of Cádiz, Spain (2,800 ha)
Hawaiian Islands, USA (2,145 ha)
Wolf Creek Dam, USA (2,000 ha)
Mai Po Marshes, Hong Kong (1,440 ha)

S
Lake Nokoué, Benin (400 ha)
Huningue, France (40 ha)
Longxian Village, China (26 ha)
Sperlonga, Italy (0.5 ha)
Paris, France (0.32 ha)

Case Study Multifunctionality

	recreational		infrastructural		ecological	
	eco-tourism programs	park integration	flood + storm management	wastewater management	integrated agriculture	RAMSAR wetlands
Sperlonga, Italy		■				
Longxian Village, China	■		■	■	■	
Comacchio, Italy	■	■	■			■
Hawaiian Islands, USA	■	■			■	
Třeboň Basin, Czech Republic	■		■			■
Pearl River Delta, China			■	■	■	
Baures, Bolivia			■		■	
Bay of Cádiz, Spain	■	■	■			■
Lake Nokoué, Benin			■		■	
Huningue, France	■	■	■			
Paris, France		■				
Kolkata, India			■	■	■	■
Mai Po Marshes, Hong Kong	■	■	■			■
Doñana Nature Reserve, Spain	■	■	■			■
Wolf Creek Dam, USA	■	■	■			

02 A scalar comparison of the case study landscapes groups them into four sizes. The multifunctionality index registers functions beyond fish farming, including recreational, infrastructural, and ecological functions. The multifunctionality of the landscape throughout its history is considered, so not all alternate functions listed here are concurrent with fish farming.

Case Study Transect Comparison

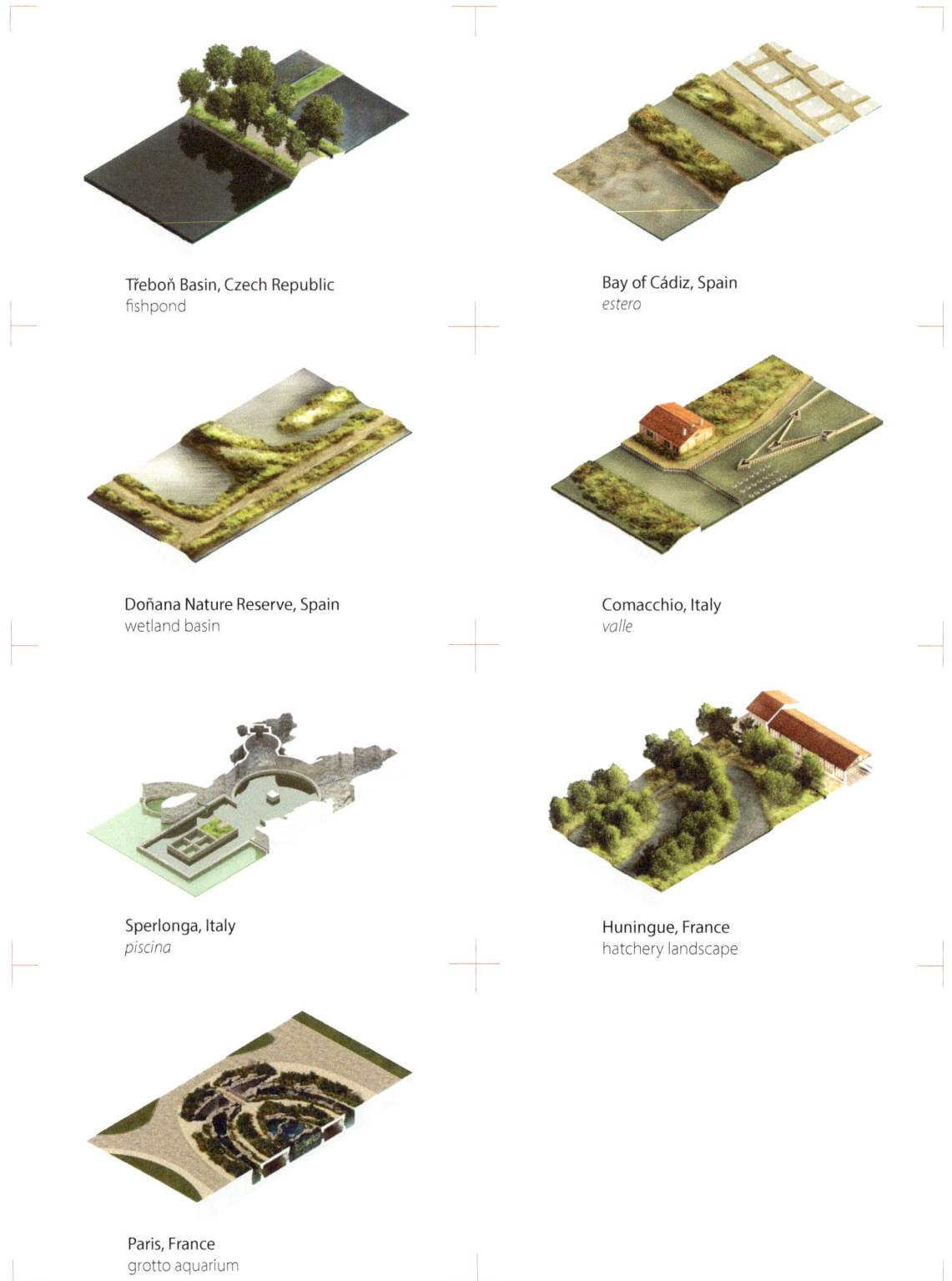

Třeboň Basin, Czech Republic
fishpond

Bay of Cádiz, Spain
estero

Doñana Nature Reserve, Spain
wetland basin

Comacchio, Italy
valle

Sperlonga, Italy
piscina

Huningue, France
hatchery landscape

Paris, France
grotto aquarium

03 Rendered axonometric projections of the fifteen case study landscapes illustrate characteristic topographic and planted forms, materiality, texture, and color. Consistency in scale, orientation, and rendering style allows for comparative analysis.

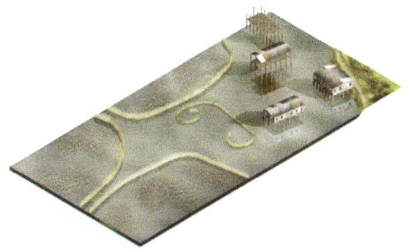

Lake Nokoué, Benin
acadja

Kolkata, India
bheri

Pearl River Delta, China
dike-pond system

Longxian Village, China
rice-fish terrace

Mai Po Marshes, Hong Kong
gei wai + fishpond

Baures, Bolivia
savanna weir

Wolf Creek Dam, USA
tailwater fishery

Hawaiian Islands, USA
ahupua'a

Case Study Water Pattern Comparison

Třeboň Basin, Czech Republic
fishpond

Bay of Cádiz, Spain
estero

Doñana Nature Reserve, Spain
wetland basin

Comacchio, Italy
valle

Sperlonga, Italy
piscina

Huningue, France
hatchery landscape

Paris, France
grotto aquarium

04 Plan diagrams of the fifteen case study landscapes. These diagrams illustrate the broad range of forms and water patterns found at aquaculture sites and systems.

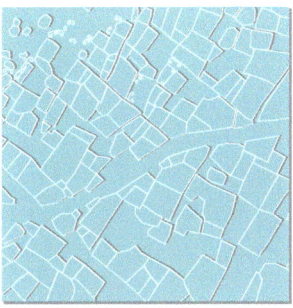

Lake Nokoué, Benin
acadja

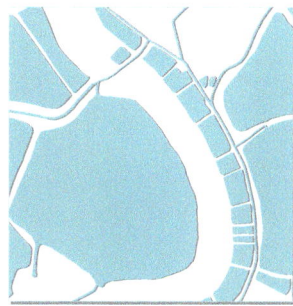

Kolkata, India
bheri

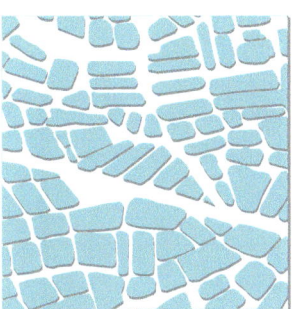

Pearl River Delta, China
dike-pond system

Longxian Village, China
rice-fish terrace

Mai Po Marshes, Hong Kong
gei wai + fishpond

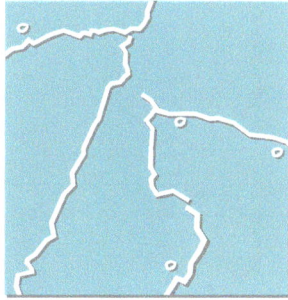

Baures, Bolivia
savanna weir

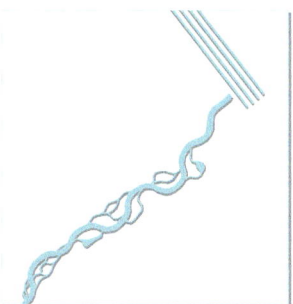

Wolf Creek Dam, USA
tailwater fishery

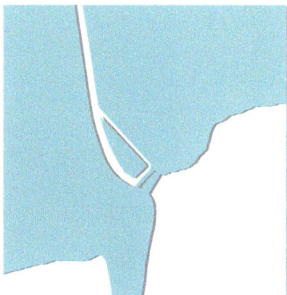

Hawaiian Islands, USA
ahupua'a

01 Active recreation along the dam at Svět pond that was constructed in the sixteenth century. An allée of pendunculate oaks (*Quercus robur*) were later planted to strengthen the earthen dam.

Fishponds of the Třeboň Basin, Czech Republic

Landscape Type:	fishpond
Landscape Area:	6,400 ha (fishponds)
	70,000 ha (Trebon PLA, 2019)
Aquaculture Yield:	1,000 kg/ha/yr
Aquaculture Type:	extensive/ semi-intensive
Water Type:	fresh

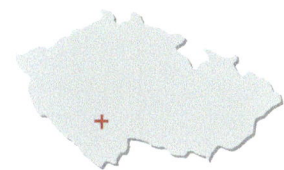

Třeboň Basin is a 70,000 ha UNESCO Biosphere Reserve and Protected Landscape Area in southern Bohemia, in the Czech Republic. This regional-scaled aquaculture landscape is characterized by nearly 500 constructed fishponds. The ponds cover 6,400 ha and are interconnected via canals that are fed and drained by the Lužnice River.[1] A royal edict issued by Emperor Charles IV in the fourteenth century expanded fishpond construction that had begun in the region in the twelfth century. Interestingly, the edict described the ponds' uses beyond fish production, including "hydrological, climate-modifying, sanitary, and aesthetic functions."[2] Construction of fishponds and canals intensified at the start of the sixteenth century, and the majority of the extant fishponds were built during this time. The Třeboň Basin fishpond system is integrated within a range of ecosystems, including peat bogs, wetlands, and extensive forests.

The contemporary Třeboň Basin fishponds are an example of a public-private partnership. The state-owned fishponds are leased to a private company that manages freshwater fish polycultures. Though the number of ponds in the region has remained constant since the nineteenth century, fish output has greatly increased due to movement toward intensive farming methods. Třeboň Basin fishponds produce nearly 3,000 mt of marketable Common Carp (*Cyprinus carpio*), making this region one of the largest suppliers of freshwater fish in Europe.[3]

Concurrent with fish farming, the Třeboň Basin fishponds support electricity generation, milling, peat production, and water-based recreation. Many of the Třeboň Basin fishponds have been listed as RAMSAR Wetlands of International Importance since 1992. However, intensive aquaculture practices have led to nutrient increases that cause eutrophication, a significant threat to the health and biodiversity of the ponds.

The Třeboň Basin is shaped by distinct hydrologic, topographic, and vegetative strategies. A system of constructed canals conveys the water from local rivers to each of the ponds. The *Zlatá stoka* (Golden Canal), a 45 km waterway constructed in 1520, remains the infrastructural linchpin of the system.[4] Valves enable periodic draining of the relatively shallow fishponds. The historic earthen dams that retain water in the ponds were originally constructed with weak local soils, so rows of oak were planted to strengthen these landforms. Today, these tree lines are an iconic feature of the landscape in this region.

Fish farming has cultural significance in the region. Fish harvests are lively public events where crowds gather to watch teams of fishermen at work in traditional garb. And in a central European tradition, farmed carp are sold live and then spend several days in household bathtubs as family pets before being served for dinner on Christmas Eve.

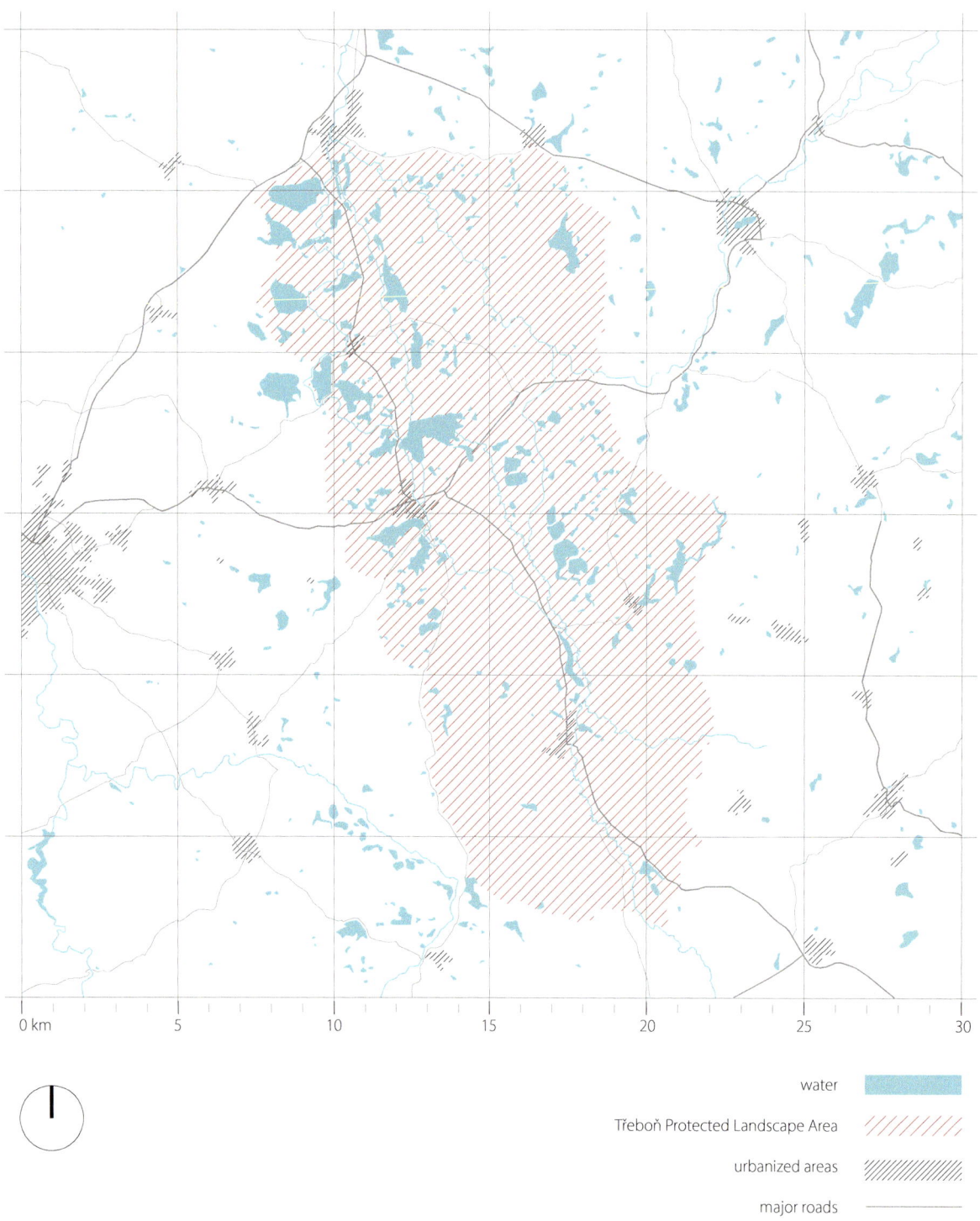

Třeboň Protected Landscape Area

urbanized areas

major roads

02 *Regional Diagram.* The Třeboň Basin fishponds cover 6,400 ha, which is 11.5% of the Třeboň Protected Landscape Area (PLA). Twenty-five thousand people live in the PLA and 9,000 live in the town of Třeboň. The enormous diversity of biotopes in the region results in high species richness of fauna, and the PLA is of particular importance as habitat for waterfowl.

water

dam

buildings

roads

trails

agriculture

03 *Site Diagram.* The contemporary plan of the Svět and Opatovický fishponds, which were constructed in the late sixteenth century near the town of Třeboň. On the landward side of the earthen dam that holds back the water of the ponds is one of the region's historical fish-holding basins. Fresh-caught fish are collected here prior to their distribution to regional markets.

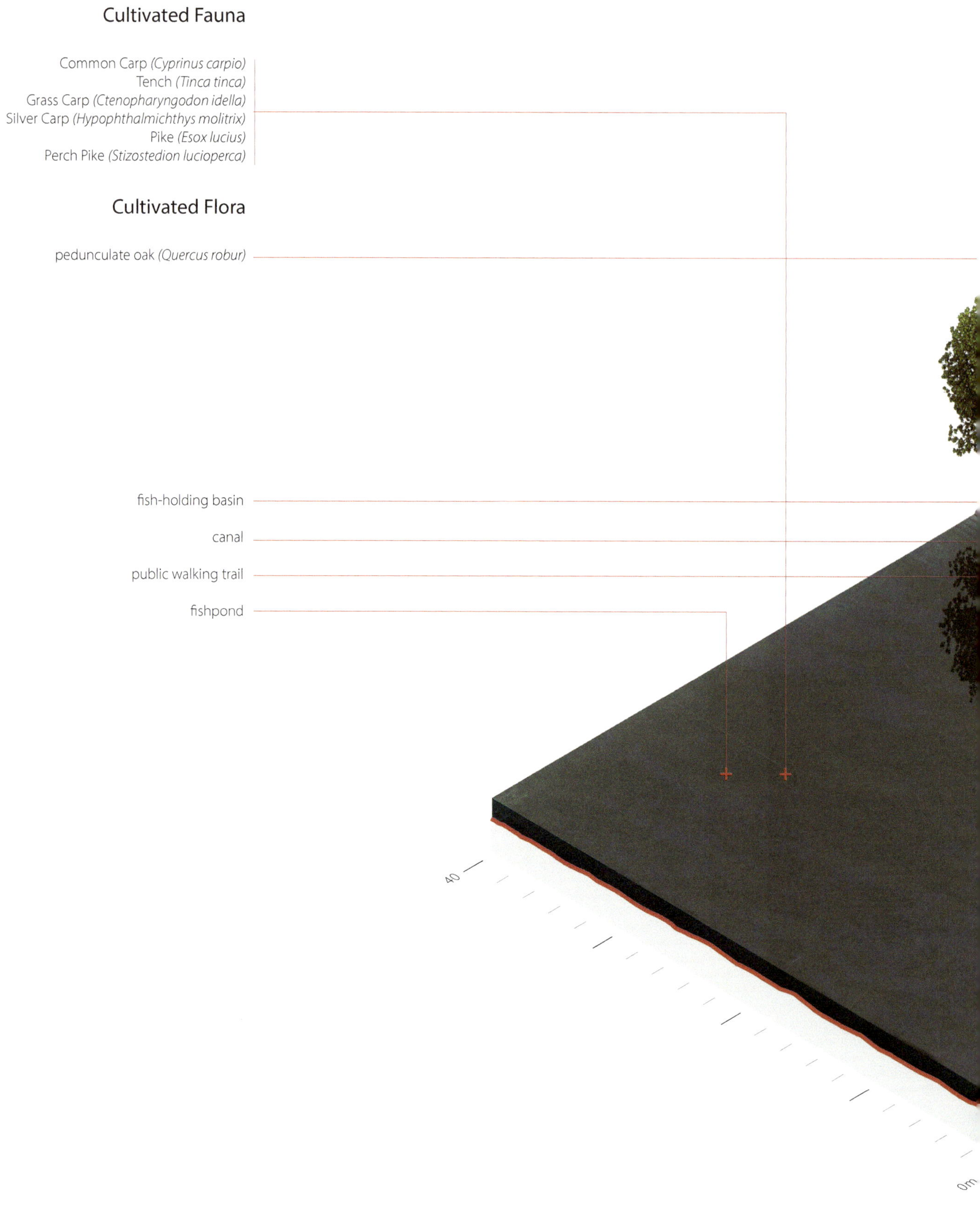

Cultivated Fauna

Common Carp *(Cyprinus carpio)*
Tench *(Tinca tinca)*
Grass Carp *(Ctenopharyngodon idella)*
Silver Carp *(Hypophthalmichthys molitrix)*
Pike *(Esox lucius)*
Perch Pike *(Stizostedion lucioperca)*

Cultivated Flora

pedunculate oak *(Quercus robur)*

fish-holding basin

canal

public walking trail

fishpond

40

0m

04 *Transect*. The earthen dams of the region double function: they retain the water of the fishponds and also serve as recreational conduits. This infrastructure is conspicuous in the landscape, due to both the height of the landform and the rows of oak trees that were planted to strengthen the dam.

05 Canal and fish-holding basins at the base of the earthen dam at Svět pond.
06 View to a row of oak trees planted at the dam at Rožmberk pond. The rows of oak trees at fishpond dams stand out from the typical pine and spruce forests that cover roughly half of the Třeboň Basin.

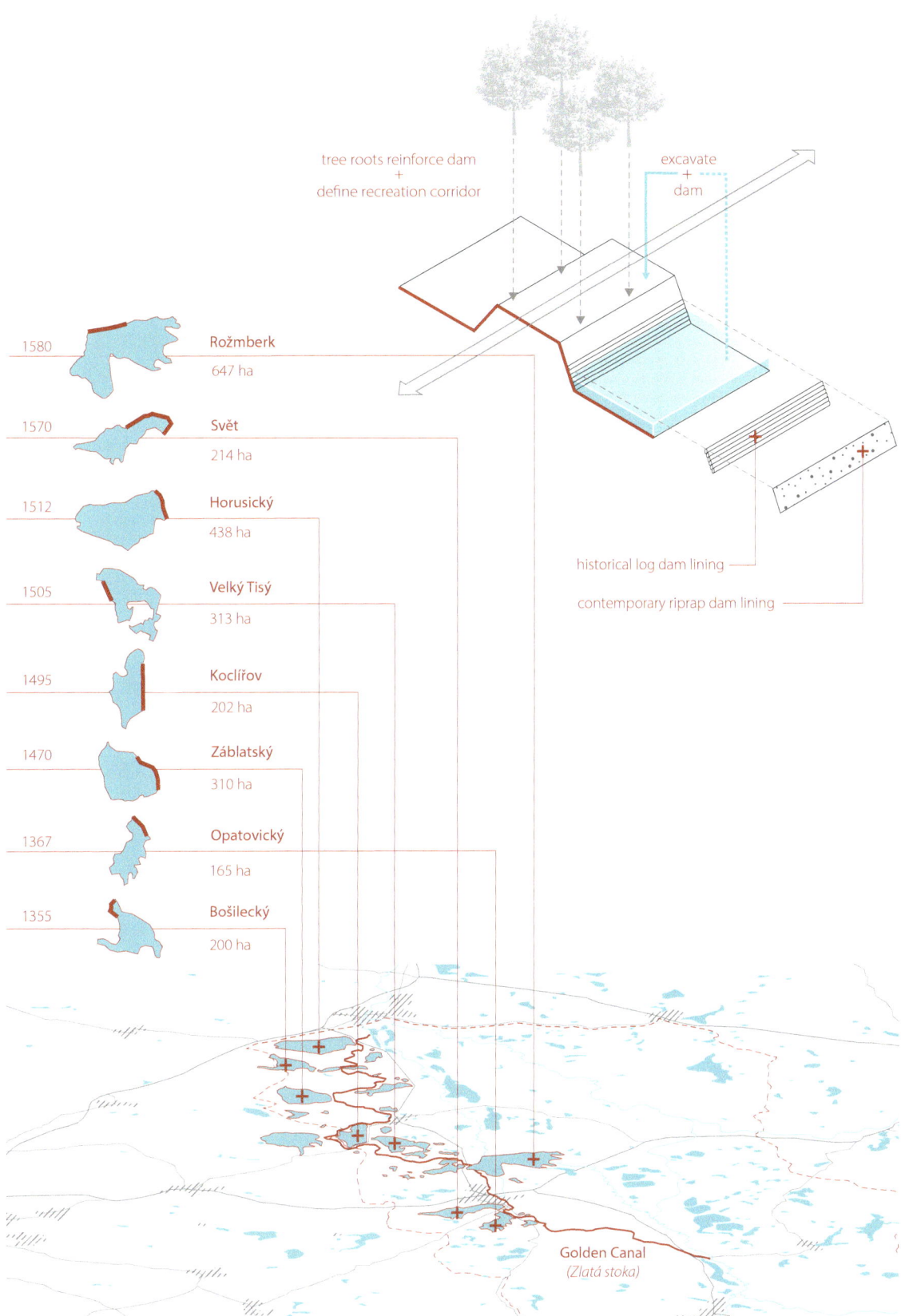

tree roots reinforce dam
+
define recreation corridor

excavate
+
dam

1580   Rožmberk
        647 ha

1570   Svět
        214 ha

1512   Horusický
        438 ha

1505   Velký Tisý
        313 ha

1495   Koclířov
        202 ha

1470   Záblatský
        310 ha

1367   Opatovický
        165 ha

1355   Bošilecký
        200 ha

historical log dam lining

contemporary riprap dam lining

Golden Canal
(Zlatá stoka)

07 *Landscape Systems and Strategies.* Fishpond construction in the region began in the fourteenth century. The Třeboň Basin fishpond system enables fish production, electricity generation, mill operations, and water-based recreation. The system also plays a critical role in floodwater management in the region.

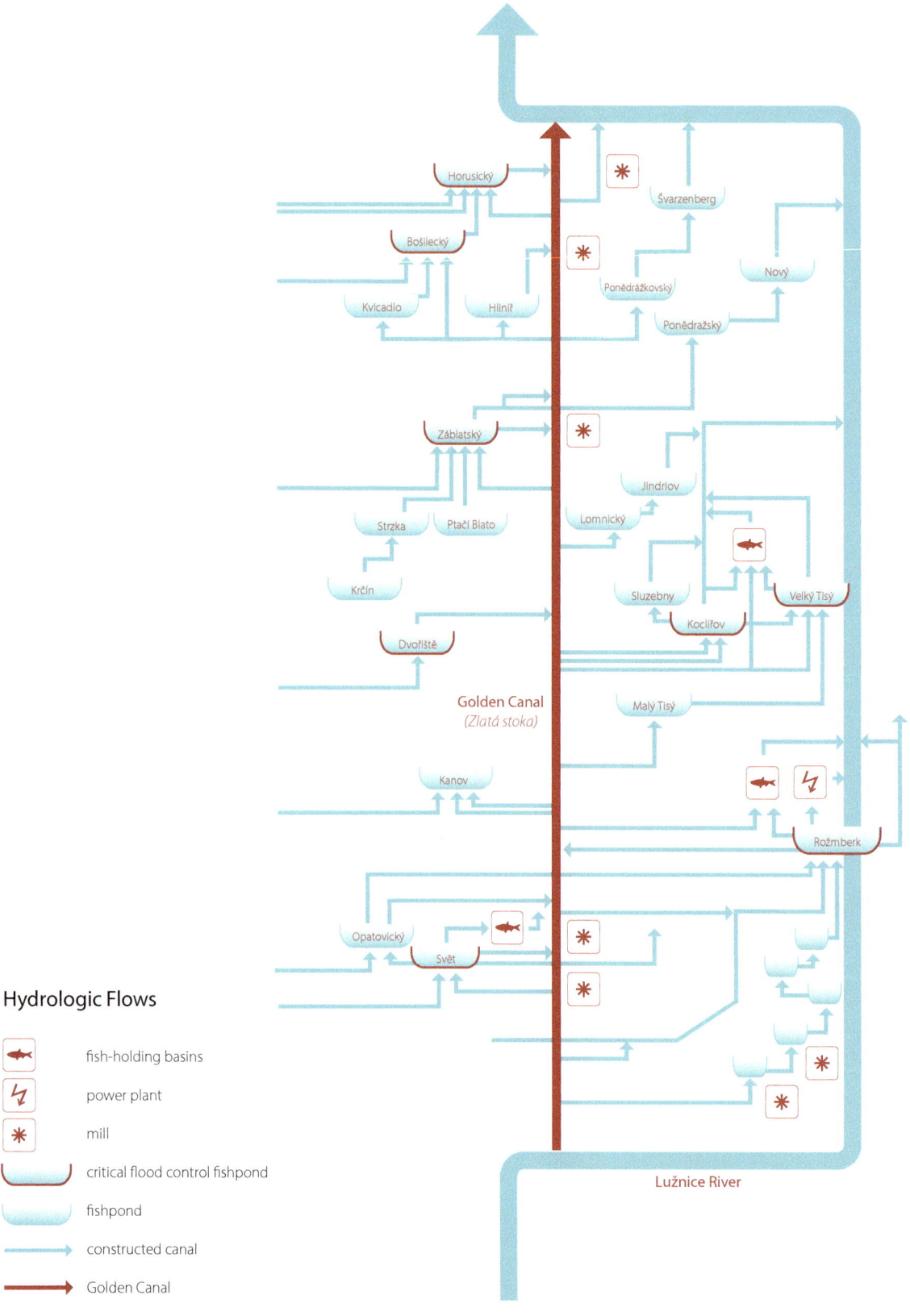

Hydrologic Flows

🐟	fish-holding basins
⚡	power plant
✳	mill
⌣	critical flood control fishpond
⌣	fishpond
→	constructed canal
→	Golden Canal

08 *Landscape Systems and Strategies.* The Golden Canal, a 45 km constructed waterway that redirects water of the Lužnice River, is a hydrologic armature that supplies and drains dozens of the fishponds. Ponds of key importance for floodwater management are identified; the retention volume of all the fishponds in Třeboň Basin is ~50–60 million m³ of water.

09 Svět pond drained for fish harvest, an activity that is undertaken by fishermen wearing traditional garb.
10 Ponds of key importance for floodwater management in the fishpond system are kept at lower than normal water levels during flood season. Sluices like this one at Rožmberk pond enable control over water level at the ponds.

01 Cyclists ride on a dike between esteros at the Salina de Carboneros, one of many salinas in the Bay of Cádiz Nature Reserve (CNR) that have transitioned into public parks and protected habitat.

Esteros in the Bay of Cádiz, Spain

Landscape Type:	esteros at salinas
Landscape Area:	2,800 ha aquaculture | 10,000 ha (Bay of Cádiz NR, 2019)
Aquaculture Yield:	600 kg/ha/yr extensive | 10-25,000 semi-extensive (2019)
Aquaculture Type:	extensive/ semi-intensive
Water Type:	brackish

For centuries, the Bay of Cádiz in Spain has been a productive saltmarsh where aquaculture and salt manufacturing are intertwined. The winding evaporation channels and basins of the *salinas* (salt farms) are constructed among tidal marshes, dunes and beaches, rivers, tidal flats, and coastal pines. These channels and basins are defining features of the landscape.

Approximately 5,500 ha of the salt marshes in the Bay of Cádiz were transformed into salinas in the eighteenth and nineteenth centuries.[1] The traditional salina includes an *estero* (large basin) that acts as a reservoir from which water is released slowly into evaporation channels for salt production. When the sluice gate of the estero opens to accept tidal waters, the basin also passively collects sea bream, mullet, and other fish species. The harvest of these interloping fish has been practiced by salina owners for centuries. Salt production at the salinas declined in the twentieth century, and some scholars note that the continuation of aquaculture practices plays a key role in maintaining the heritage value of the landscape.[2]

The annual harvests of fish in esteros, called *despesques,* are embodied encounters between humans and fish. These events draw crowds of enthusiasts who either watch as farmhands gather hundreds of thrashing fish using a seine net or enter the shallow waters themselves to catch fish in hand held nets. More solitary fish harvesting practices are also prevalent in the area, including *marisqueo,* the gathering of shellfish by hand in shallow waters, as well as recreational angling.

In 1989, the 10,000 ha Bay of Cádiz Nature Reserve (CNR) was established and ecotourism infrastructure was constructed, including an education center, extensive trails, and bird blinds. The CNR is nested within a polynuclear urban agglomeration, comprised of five municipalities with a total population of more than 400,000. Despite this urban context it remains an important site for migrating and wintering water birds. It is a refuge for the breeding populations of five different shorebird species. The Bay of Cádiz has been designated a RAMSAR Wetland of International Importance since 2003.

At the turn of the twenty-first century, fish were cultivated and harvested in approximately 2,000 ha of esteros, and aquaculture practices in the Bay of Cádiz have increased as the winding channels of salinas are transformed into repetitive basins for semi-intensive aquaculture.[3] The Province of Cádiz contributes to more than 50% of the total aquaculture production of Andalusia, with most of the fish originating from farms in the Bay of Cádiz.

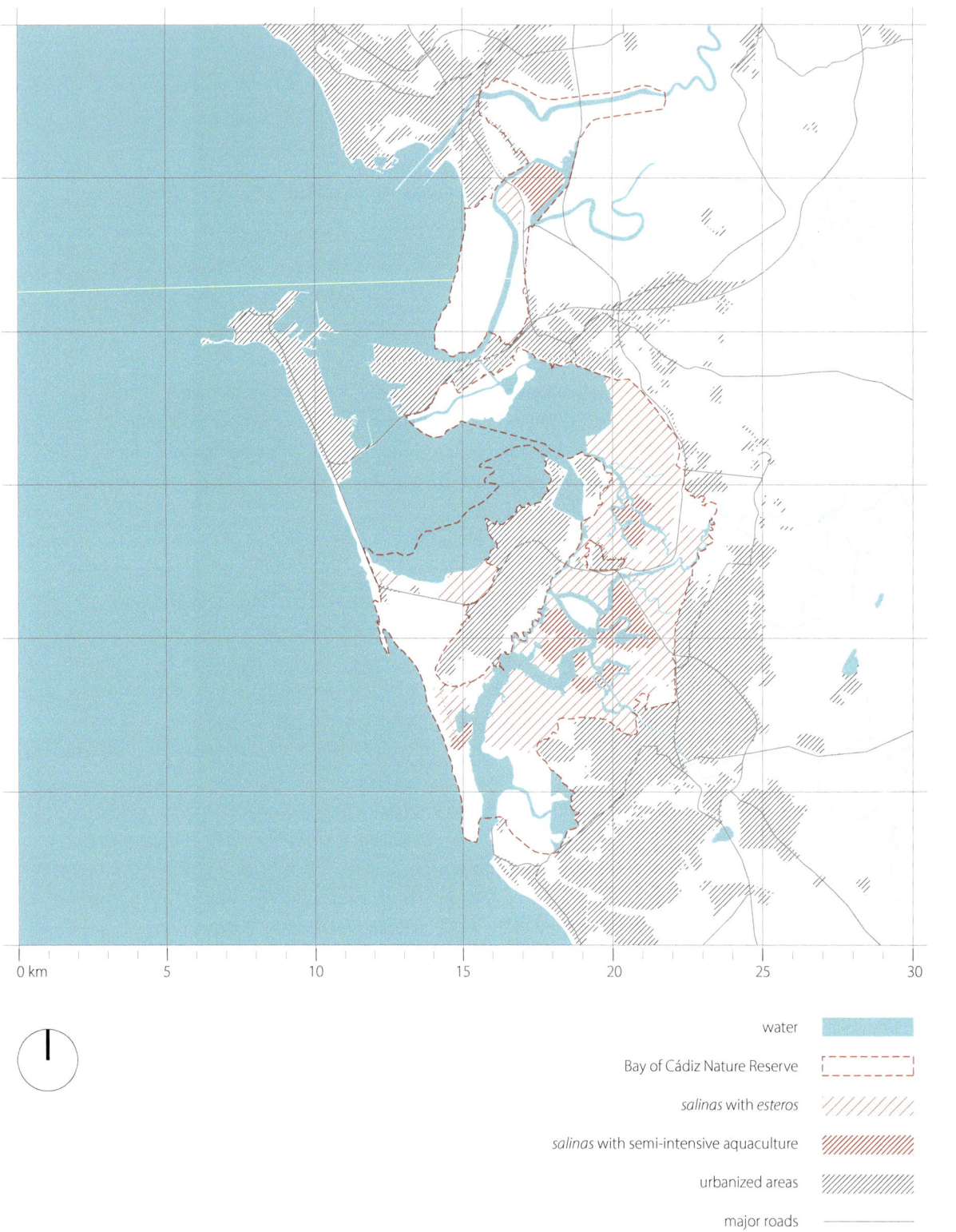

水 water

Bay of Cádiz Nature Reserve

salinas with *esteros*

salinas with semi-intensive aquaculture

urbanized areas

major roads

0 km   5   10   15   20   25   30

02 *Regional Diagram*. The 10,000 ha CNR is nested within a polynuclear urban agglomeration comprised of five municipalities: San Fernando, Chiclana de la Frontera, Puerto Real, Puerto de Santa María, and Cádiz. The reserve is characterized by historical salt and fish production landscapes.

water

buildings

roads

trails

wetland / open space

03 *Site Diagram.* At Salina de San Carlos y San Jaime, the byzantine troughs of the salina have been transformed into regimented basins for semi-intensive fish production. Topographic and hydrologic adjustments to facilitate new aquaculture practices is increasing at salinas across the CNR.

Cultivated Fauna

Sole *(Solea senegalensis)*
Gilthead Seabream *(Sparus aurata)*
Grey Mullet *(Mugil cephalus)*
Sea Bass *(Dicentrarchus labrax)*
European Eel *(Anguilla anguilla)*
Manila Clam *(Ruditapes phllippinarum)*
Portuguese Oyster *(Crassostrea angulata)*

Ambient Flora

sea lettuce *(Ulva lactuca)*
sea grass *(Ruppia cirrhosa)*
cordgrass *(Spartina spp.)*
glasswort *(Salicornia spp.)*
Mediterranean statice *(Limonium sinuatum)*
mastic tree *(Pistacia lentiscus)*
European fan palm *(Chamaerops humilis)*

tageria (salt crystalization basins)

veluta (evaporation canals)

| 80

| 40

estero

height of water in fully filled *estero*

04 *Transect.* Basins and canals of depths ranging from 1.5 m to just a few centimeters are constructed at salinas for salt and fish production. The deepest basin of a salina, the estero, serves as a reservoir for salt production processes while also serving as habitat for fish and water birds. Oyster shells are often used to reinforce the dikes of salinas in this area.

05 Tourists and locals observe the annual fish harvest at the estero at Salina de Chiclana.

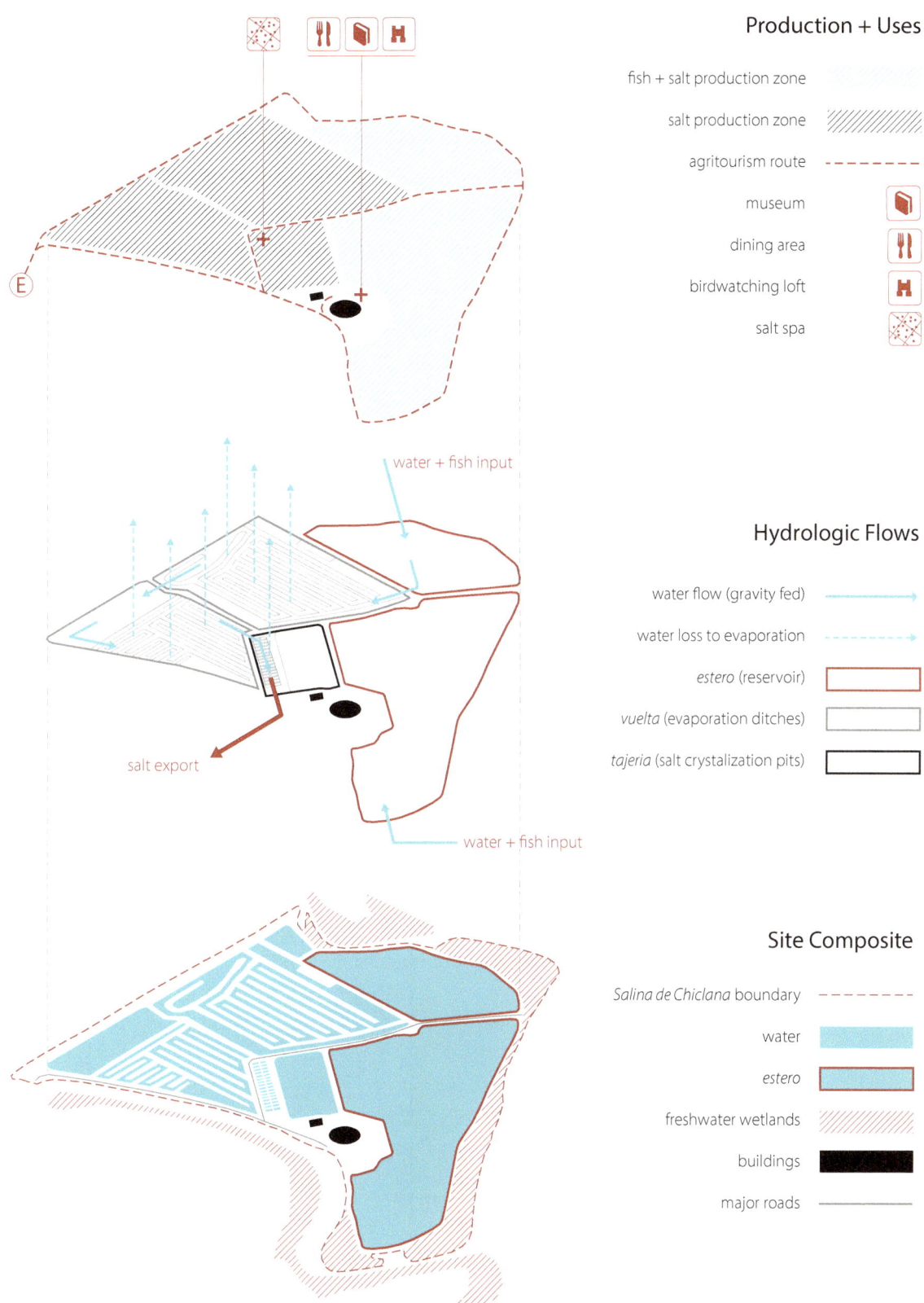

Production + Uses

fish + salt production zone	
salt production zone	
agritourism route	– – – – –
museum	📖
dining area	🍴
birdwatching loft	Ħ
salt spa	▨

Hydrologic Flows

water + fish input

salt export

water + fish input

water flow (gravity fed)	→
water loss to evaporation	⇢
estero (reservoir)	
vuelta (evaporation ditches)	
tajeria (salt crystalization pits)	

Site Composite

Salina de Chiclana boundary	– – – – –
water	
estero	
freshwater wetlands	
buildings	
major roads	——

06 *Landscape Systems and Strategies.* At Salina de Chiclana the landscape is designed for artisanal-scale salt production activities and fish production. Ecotourism infrastructure includes biking trails, bird blinds, a museum, and a salt spa.

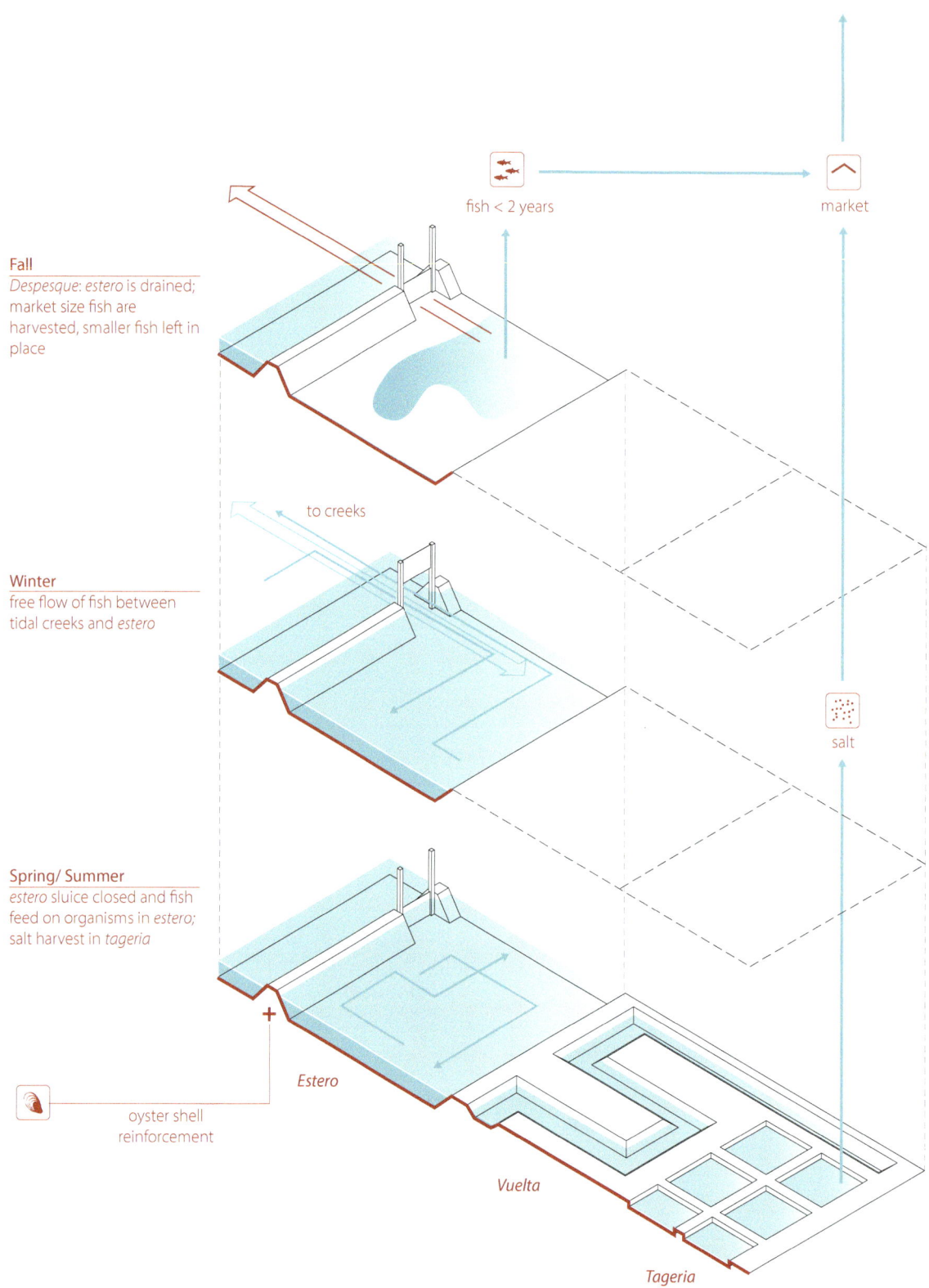

Fall
Despesque: estero is drained; market size fish are harvested, smaller fish left in place

fish < 2 years

market

to creeks

Winter
free flow of fish between tidal creeks and *estero*

salt

Spring/ Summer
estero sluice closed and fish feed on organisms in *estero*; salt harvest in *tageria*

oyster shell reinforcement

Estero

Vuelta

Tageria

07 *Landscape Systems and Strategies.* Esteros used for fish cultivation are found throughout the patchwork of basins, canals, and tidal creeks at the CNR. Fish feed and grow in esteros as the salt harvest occurs in the spring and summer, and market-size fish are harvested in the fall.

08 Recreational fishing at an estero at Salina de Carboneros.
09 Salt spa at the tageria (salt crystallization pits) at Salina de Chiclana.

01 Greater flamingos (*Phoenicopterus ruber*) foraging in wetland basin. Beyond this wetland basin are nursery ponds that are covered in nets to protect the small fish from avian predation.

Wetland Basins in the Doñana Nature Reserve, Spain

Landscape Type: wetland basin

Landscape Area: 3,200 ha (wetland basin) | 11,000 ha (Veta la Palma)

Aquaculture Yield: 470 kg/ha/yr

Aquaculture Type: extensive/ semi-intensive

Water Type: brackish

Veta la Palma is an 11,000 ha commercial fishery located within the Doñana Nature Reserve, a protected fluvial-coastal system of natural salt marshes, beaches, and forests centered at the delta of the Guadalquivir River in southern Spain. The constructed wetland basins at Veta la Palma support extensive and semi-intensive aquaculture as well as a significant range of ecosystem services including water quality improvement, promotion of biodiversity, and ecotourism.

Beginning in the early twentieth century, marshes in this area were drained to support a rice monoculture. In 1982, the private company Pesquerías Isla Mayor, S.A. purchased the land to establish an aquaculture operation, reversing the water flow of a 300 km in length network of canals on the site. The drainage canals became irrigation channels. Today the brackish water of the Guadalquivir River inundates the channels and fills the array of basins that serve as aqueous habitat and fishponds. In contrast to the Doñana Park system, with its long dry seasons, Veta la Palma is managed as a wet landscape year-round and is a RAMSAR Wetland of International Importance. As a public-private partnership, Veta la Palma serves as a model for the progressive management of other degraded coastal wetlands in the Mediterranean area.[1]

Miguel Medialdea, the Quality and Environment Manager of Veta la Palma, describes the farm as, "A mosaic of habitats defined by variations in soil conditions and humidity. . . . This carefully managed artificial wetland supports a rich and nourished flora and fauna, including dense communities of microalgae and invertebrate species, which are the foundation of our aquaculture."[2] Fish and birds consume the shrimp and microalgae that flourish in this habitat. Microalgae also play a role in cleansing the water of organic nutrients due to both high volumes of fish waste and fertilizer runoff from upriver agricultural practices.[3] The species of fish that are farmed for market are predominantly sea bass and mullet, and the annual fish production of approximately 1,400 mt is sold to gourmet-food shops and haute-cuisine chefs in Mediterranean countries and the United States.

The flooded site is also an attractive stop-over for migratory birds, traveling from northern Europe to the African continent. Over 250 bird species routinely visit the site to nest in the variety of habitat islands constructed within the floodable basins.[4] Sediment that is periodically dredged from basins to maintain adequate depths for aquaculture is then utilized to create the island chains. The wild birds and farmed fish at Veta la Palma also underpin a recreational itinerary that includes bird watching, participation in fish harvests, and horse-drawn boat tours.

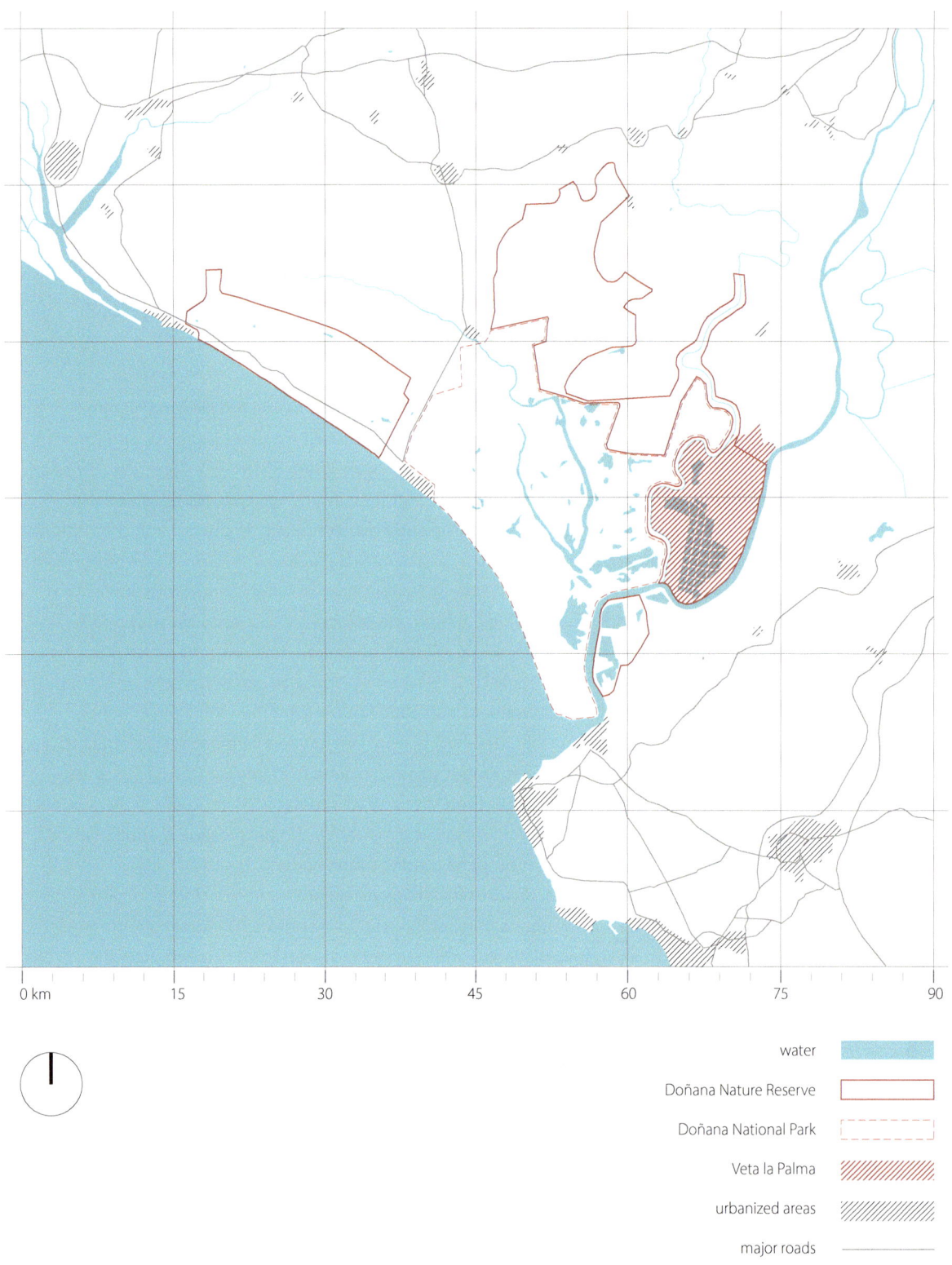

water ▨

Doñana Nature Reserve ▭

Doñana National Park ▭

Veta la Palma ▨

urbanized areas ▨

major roads ——

0 km	15	30	45	60	75	90

02 *Regional Diagram.* Veta la Palma is a commercial fish farm constructed on an island in the delta of the Guadalquivir River. The farm falls within the boundary of the Doñana Nature Reserve, a protected zone characterized by coastal dunes, salt marshes, and pine forests. Rice and salt farms are prevalent in the region outside the boundaries of the reserve.

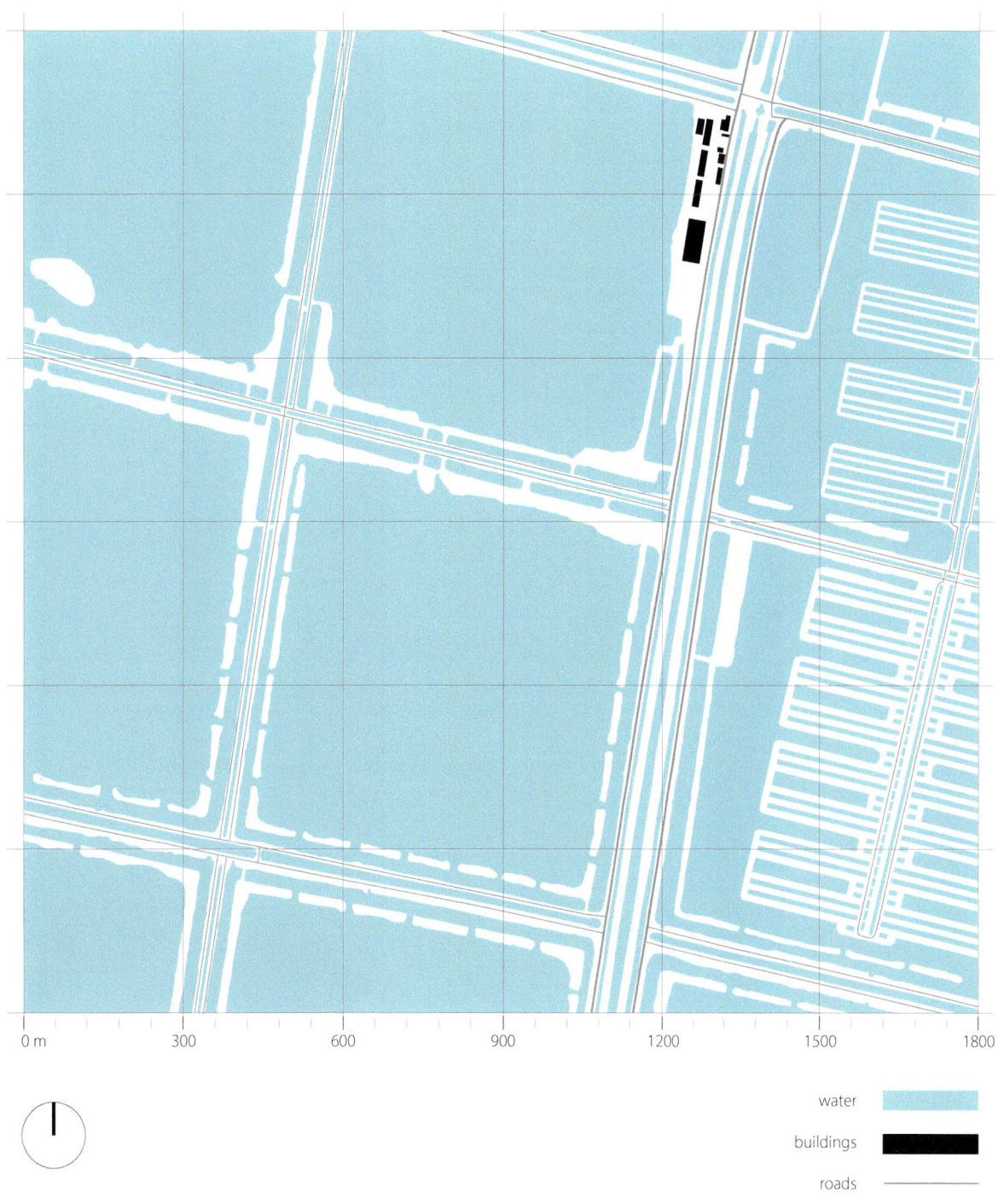

| 0 m | 300 | 600 | 900 | 1200 | 1500 | 1800 |

water

buildings

roads

03 *Site Diagram*. At Veta la Palma fish grow out for three years in the large, square wetland basins. The fish forage for microalgae and shrimp that are drawn into the system along with water from the Guadalquivir River estuary. Some wetland basins are lined with habitat islands that provide nesting areas for waterfowl.

Cultivated Fauna

Sea Bass *(Dicentrarchus labrax)*
Flathead Grey Mullet *(Mugil cephalus)*
Thinlip Mullet *(Mugil ramada)*
Gilthead Seabream *(Sparus aurata)*
European Eel *(Anguilla anguilla)*
Meagre *(Argyrosomus regius)*
Sole *(Solea senegalensis)*

cattle *(Bos Taurus)*
horse *(Equus spp.)*

Ambient Flora

seablite *(Suaeda spp.)*
cordgrass *(Spartina spp.)*

extensive culture basin

constructed habitat island

access road

water supply canal

semi-intensive grow out basin

40

0m

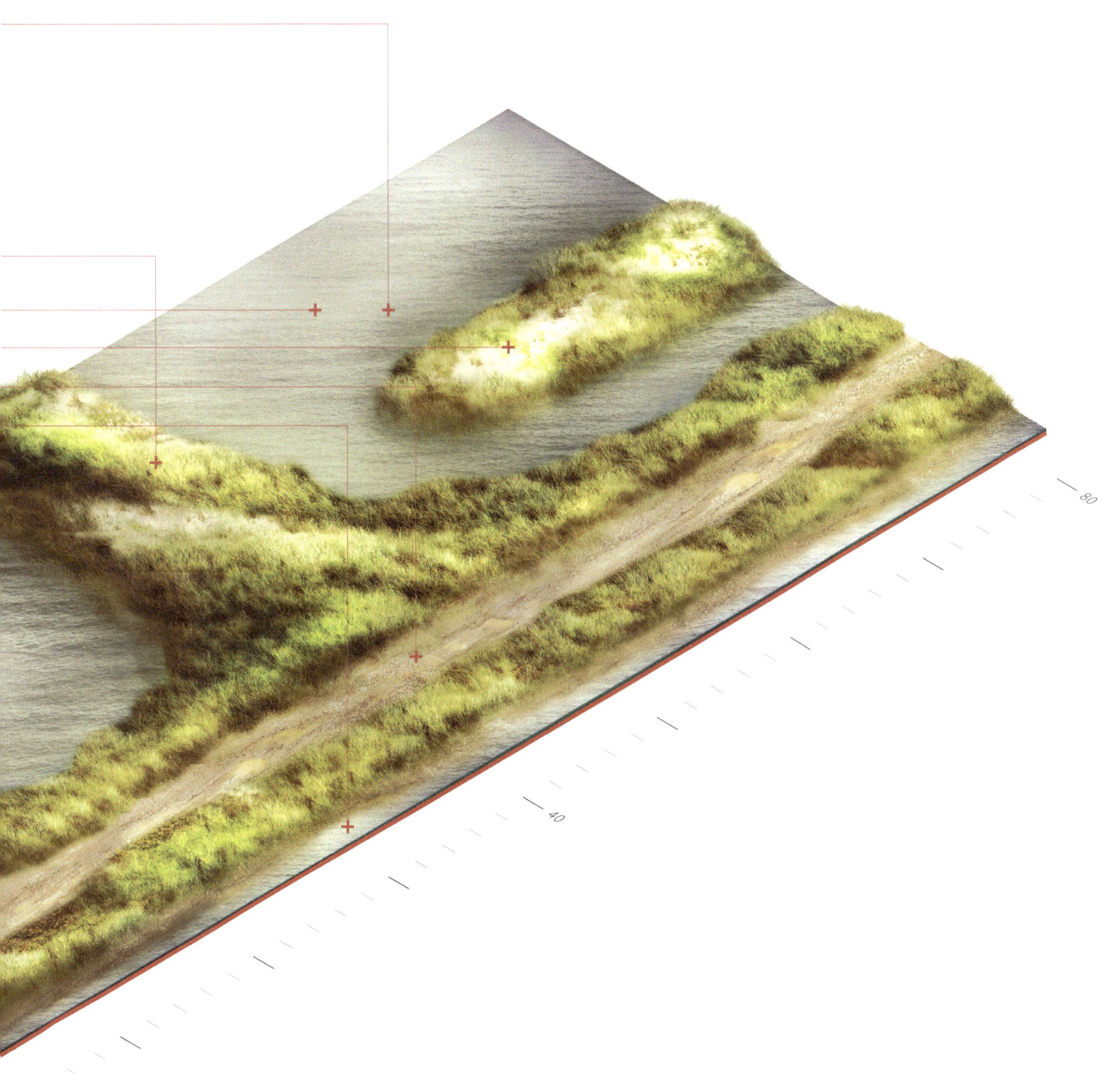

04 *Transect.* Veta la Palma's 3,200 ha fish farming area is divided into 45 rectangular, 70 ha interconnected basins. The average depth of the basins is 0.5 meters. Lining the edges of these large, shallow basins are narrow basins for semi-intensive fish production, earthen canals, and constructed habitat islands.

05 Fishing with a seine net at a wetland basin.
06 Ecotourists are pulled across a wetland basin in a horse-drawn *patera*.

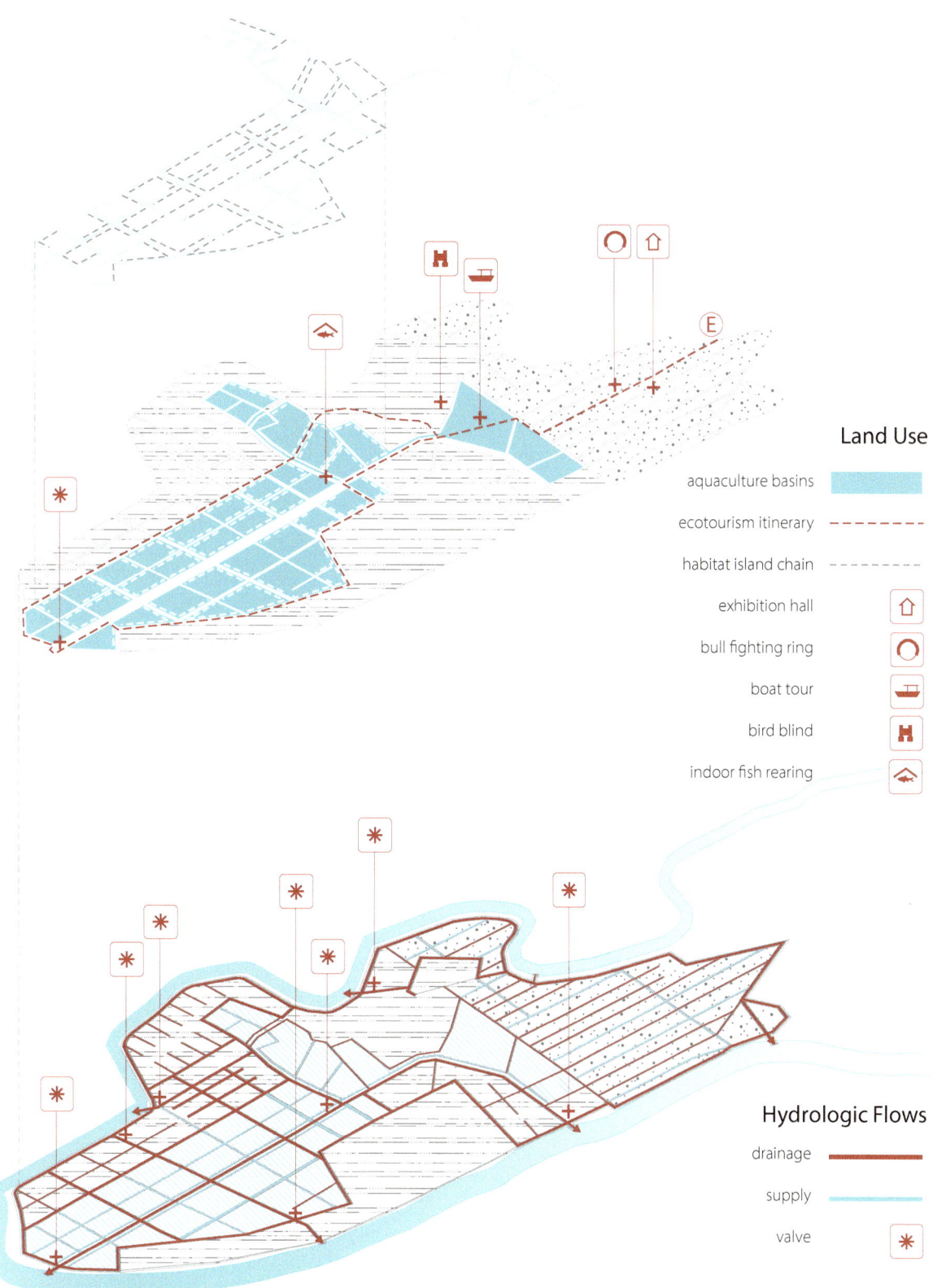

Land Use

aquaculture basins

ecotourism itinerary   – – – – – –

habitat island chain   – – – – – –

exhibition hall   ⬆

bull fighting ring   ◯

boat tour   ⛴

bird blind   ⓗ

indoor fish rearing   🐟

Hydrologic Flows

drainage   ▬▬▬

supply   ▬▬▬

valve   ✳

Guadalquivir River

07 *Landscape Systems + Strategies.* A series of valves facilitate supply and drainage of water from the Guadalquivir River to wetland basins, conservation areas, and sustainable agriculture for livestock feed. One million cubic meters of water is pumped daily from the river along 300 km of channels.

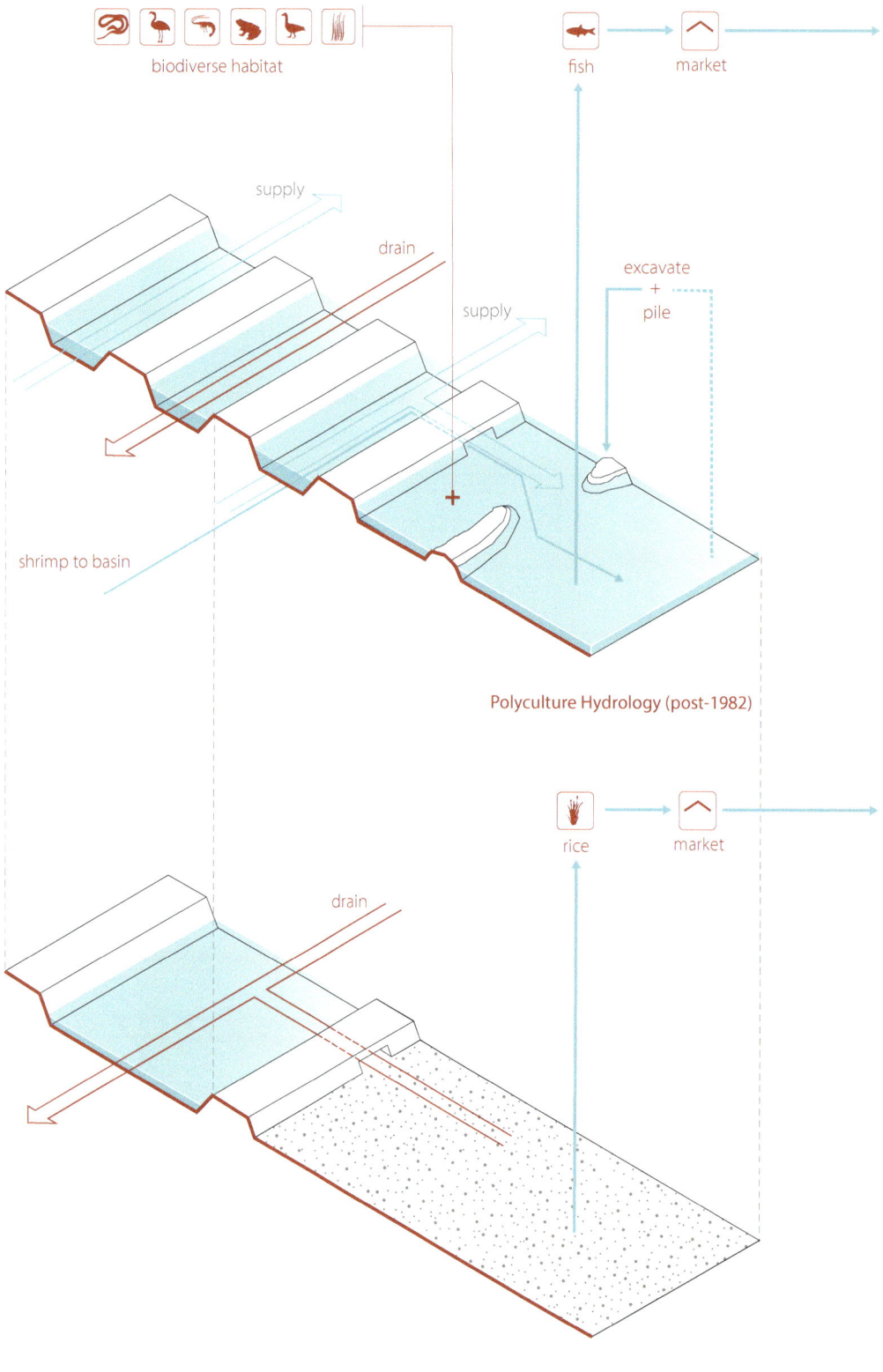

biodiverse habitat

fish

market

supply

drain

supply

shrimp to basin

excavate
+
pile

Polyculture Hydrology (post-1982)

rice

market

drain

Monoculture Hydrology (pre-1982)

08 *Landscape Systems and Strategies.* Reversing the flow of water across Veta la Palma from drainage to irrigation enabled a pivot from a rice monoculture to a fish polyculture. The wetland basins are used for commercial aquaculture, but they also serve as refuge for endangered fish species. Over 100 habitat islands lining the basins were created for nesting waterfowl.

09 Sluice at irrigation canal that supplies water to wetland basins.
10 Flamingos take flight next to a habitat island.

01 Aerial view of the reconstructed lavoriero at the Stazione di Pesca Serilla. The restored buildings date to the eighteenth century and originally functioned as equipment storage and housing for workers during the harvest season.

Valli of Comacchio, Italy

Landscape Type:	valle
Landscape Area:	10,000 ha (2019) | 49,000 ha (19th C)
Aquaculture Yield:	16.4 +/- 6.5 kg/ha/yr
Aquaculture Type:	extensive/ semi-intensive
Water Type:	brackish

For centuries, coastal lagoons on the Italian peninsula have been sites of increasingly complex topographic and hydrologic adaptations for aquaculture. The Valli di Comacchio, located along the Italian coast of the Adriatic Sea, exemplifies the transformation of coastal lagoons into aquaculture landscapes. The term *valle* (*valli*, plural) describes an area of coastal lagoon that is enclosed by embankments and features canals with elaborate barriers used to capture European Eel (*Anguilla anguilla*), as well as sea bass and mullet that migrate between the valle and the sea.[1] Beginning in the thirteenth century, individual management of the valli evolved into an extensive cooperative and social system that eventually included over four hundred public and private farms.[2]

The Valli di Comacchio lagoon complex comprised 49,000 ha at its largest size in the late nineteenth and early twentieth centuries. At the Valli di Comacchio, a branching system of canals formed by earthen dikes creates a conduit from the Adriatic Sea to the inland valli. The canals allow the warm, oxygen-rich sea water to enter the brackish valli and provide a route for migratory eels and fish that live in the valli but spawn in the sea. *Lavorieri* (*lavoriero*, singular), complex triangular traps constructed with wood and reed screens, are erected at the nexus of the canal and the valli to capture the aquatic fauna, swimming toward the sea.[3] The main fishing season runs from mid-September to mid-December, a time when water levels in the valli are rising as sea water enters the lagoon. A second, shorter catch occurs during early spring until April. Beginning in the twentieth century, land reclamation for field agriculture reduced the size of the Valli di Comacchio to its present size of 10,000 ha. Only four lavorieri are still in operation and they are owned and managed by the Town of Comacchio. The historic Manifattura dei Marinati facility in the Town of Comacchio, with its impressive Sala dei Fuochi (Hall of Fires) that features twelve large fireplaces, continues centuries-old traditions of roasting and marinating eels.

Water in the valli does not freeze in winter due to its high salinity and this makes the lagoon an optimal winter resting ground for more than 300 bird species.[4] The rich variety of habitats and high biodiversity of the lagoon led to the designation of the Valli di Comacchio as a RAMSAR Wetland of International Importance in 1976. Today the Valli di Comacchio is one of the stations in the Regional Park of the Po Delta and revenue from ecotourism and recreation is increasing. Extant and reconstructed lavorieri are accessible to tourists by trail and boat. In addition, a new form of fishing infrastructure has emerged at the Valli di Comacchio; recreational fishing lodges with massive operable nets that can be gently lowered into the water to scoop up sport fish. Known locally as *bilancioni*, the lodges line the canals and can be leased by citizens from the state.

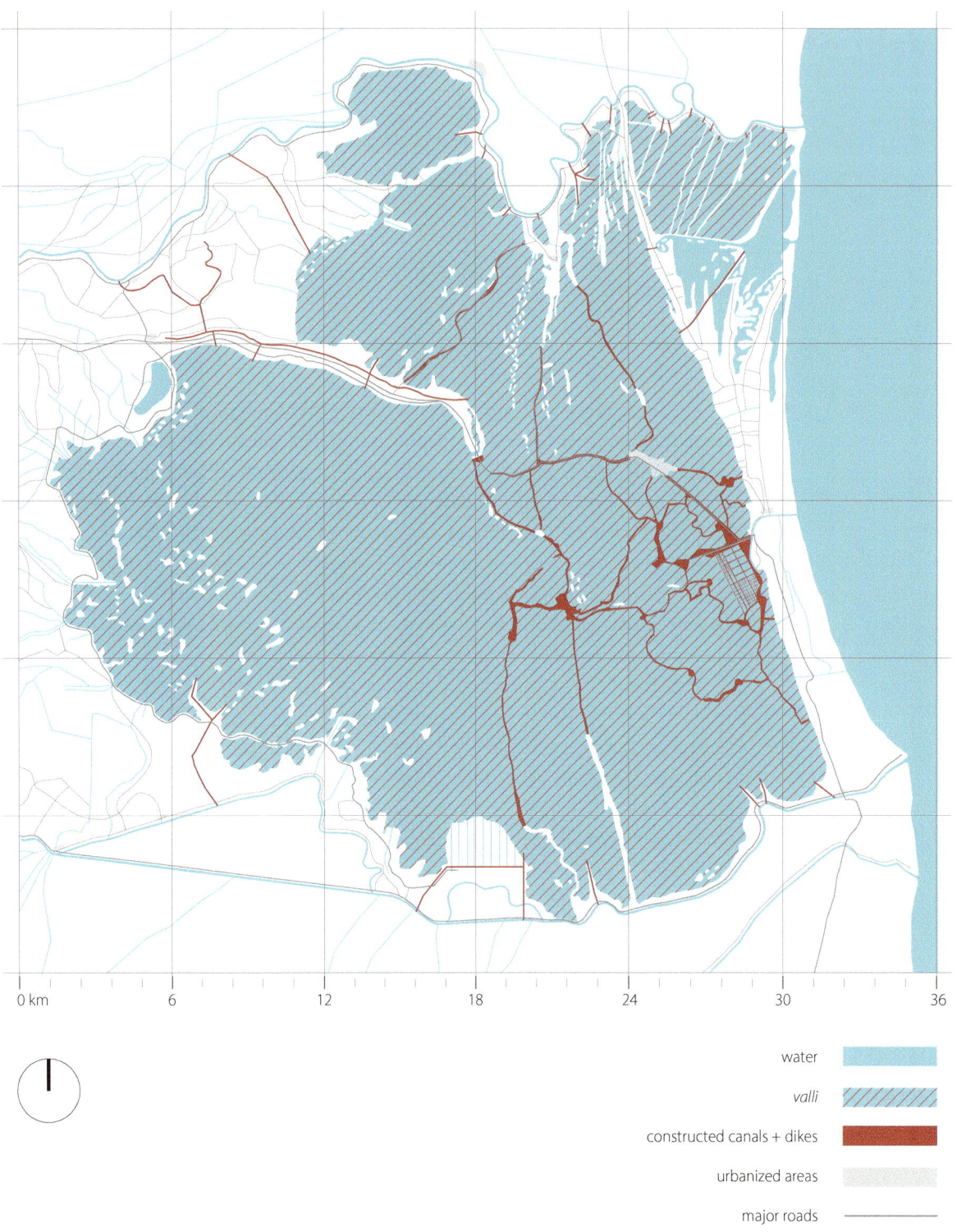

water

valli

constructed canals + dikes

urbanized areas

major roads

0 km   6   12   18   24   30   36

02 *Regional Diagram*. The Valli di Comacchio, depicted here at the turn of the twentieth century, was a ~49,000 ha brackish coastal lagoon that was subdivided into dozens of valli to facilitate the harvest of eels. The Town of Comacchio resides within the valli, as does a large salt production operation established in the mid-nineteenth century.

water			
salt water canals			
lavoriero			
buildings			
agriculture			

03 *Site Diagram.* Aquaculture and salt production are colocated at the Valli di Comacchio. An extensive network of salt water canals supply both operations. The canals are formative elements in the urban fabric of the Town of Comacchio as well. Many of the canals that feed the valli are capped by lavorieri that were constructed to capture and harvest eels and fish.

Cultivated Fauna

European Eel *(Anguilla anguilla)*
mullet *(Mugil and Liza spp.)*
Sea Bass *(Dicentrarchus labrax)*
Sea Bream *(Sparus aurata)*
Plaice *(Platichthys flesus luscus)*
Silverside *(Atherina boyeri)*

Ambient Flora

sea grass *(Ruppia cirrhosa)*
sea blite *(Suaeda maritima)*
hairy goosefoot *(Bassia irsuta)*
rooting lead-grass *(Arthrocnemum perenne)*
sea purslane *(Halimione portulocoides)*
sea lavender *(Limonium serotinum)*
sea starwort *(Aster tripolium)*

casone

covole (canal)

lavoriero

sluice

bolaghe (live eel storage containers)

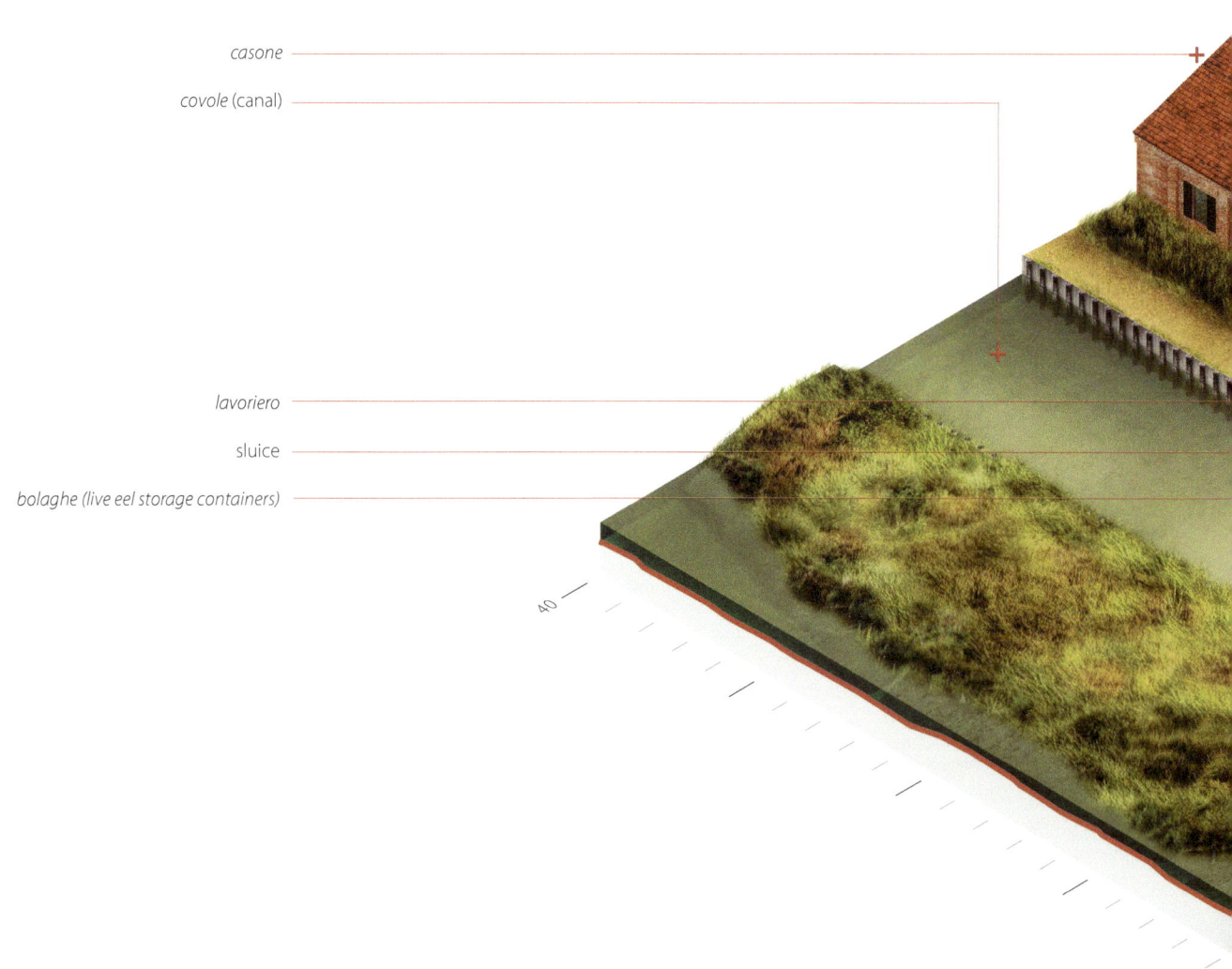

40

0m

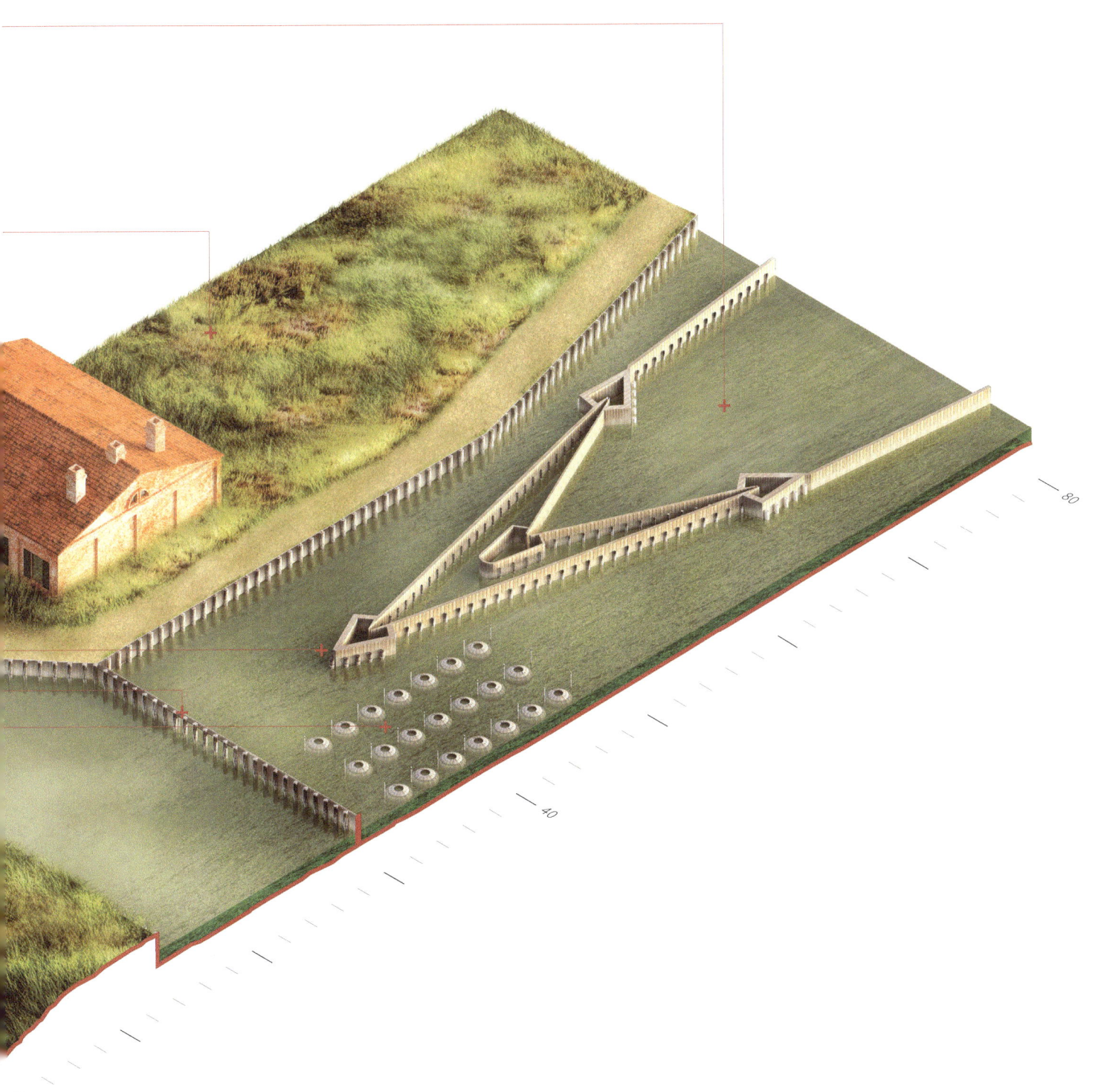

04 *Transect*. A lavoriero is an eel trap that was historically constructed with wood piles and reed screens. The lavorieri were constructed to capture and harvest eels and fish that would migrate from the valli to the Atlantic Ocean to spawn. Casoni, buildings housing workers and equipment, were located adjacent to many of the lavorieri at the Valli di Comacchio.

05 The Lavoriero di Serilla is reconstructed using traditional materials.
06 Evaporation basins at a reconstructed portion of the nineteenth century Salina di Comacchio.
07 Bird watchers can observe over three hundred bird species here, including waterfowl, waders, gulls, and terns.

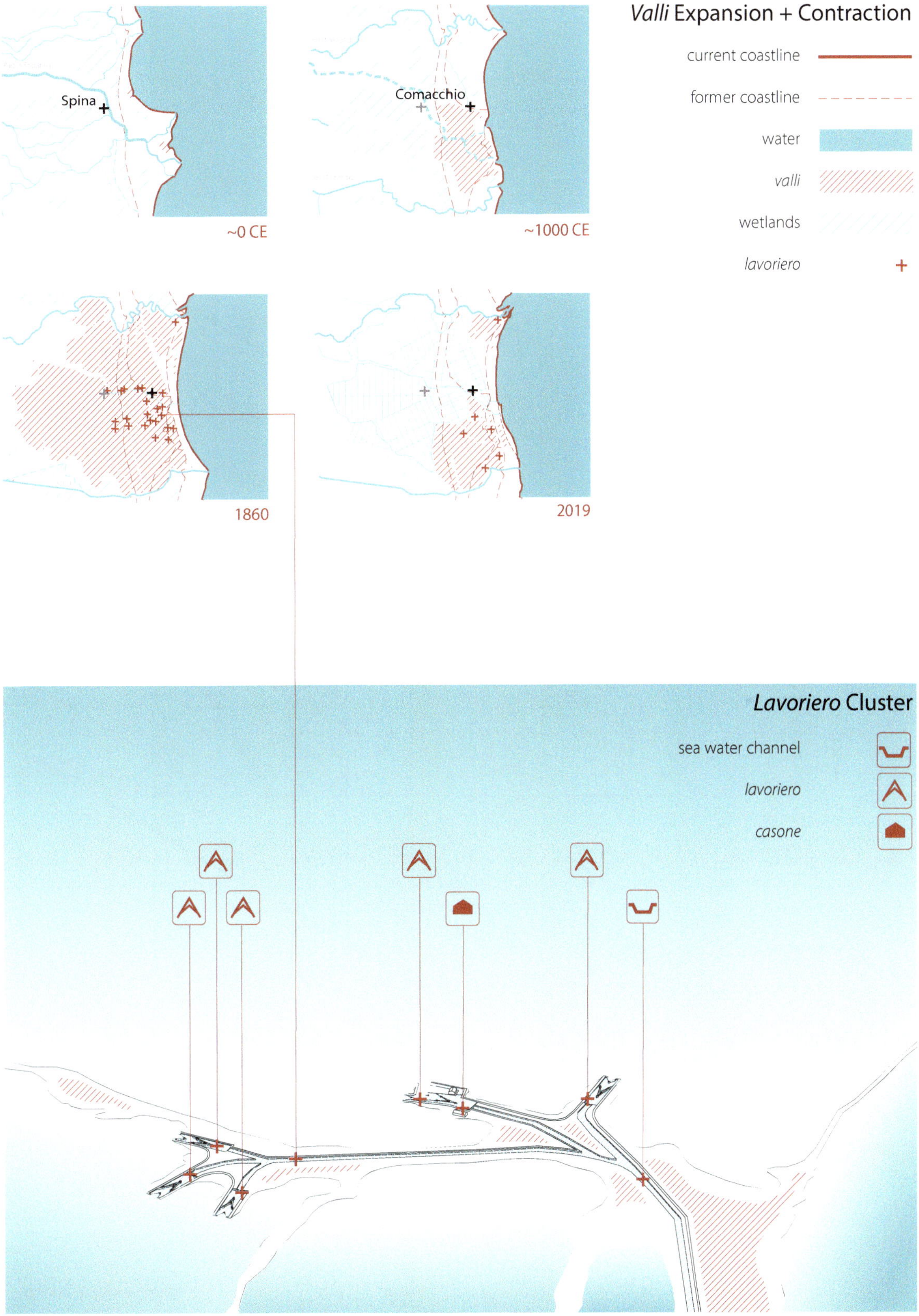
Valli Expansion + Contraction
current coastline
former coastline
water
valli
wetlands
lavoriero
Spina
~0 CE
Comacchio
~1000 CE
1860
2019
Lavoriero Cluster
sea water channel
lavoriero
casone

08 *Landscape Systems and Strategies*. Subsidence at the swamps of the fan delta of the ancient Po River led to the intrusion of sea water. Construction of dikes and canals for eel and fish harvest likely began around 1,000 CE. By the mid-nineteenth century the valli system was at its most extensive, but its size reduced substantially in the twentieth century.

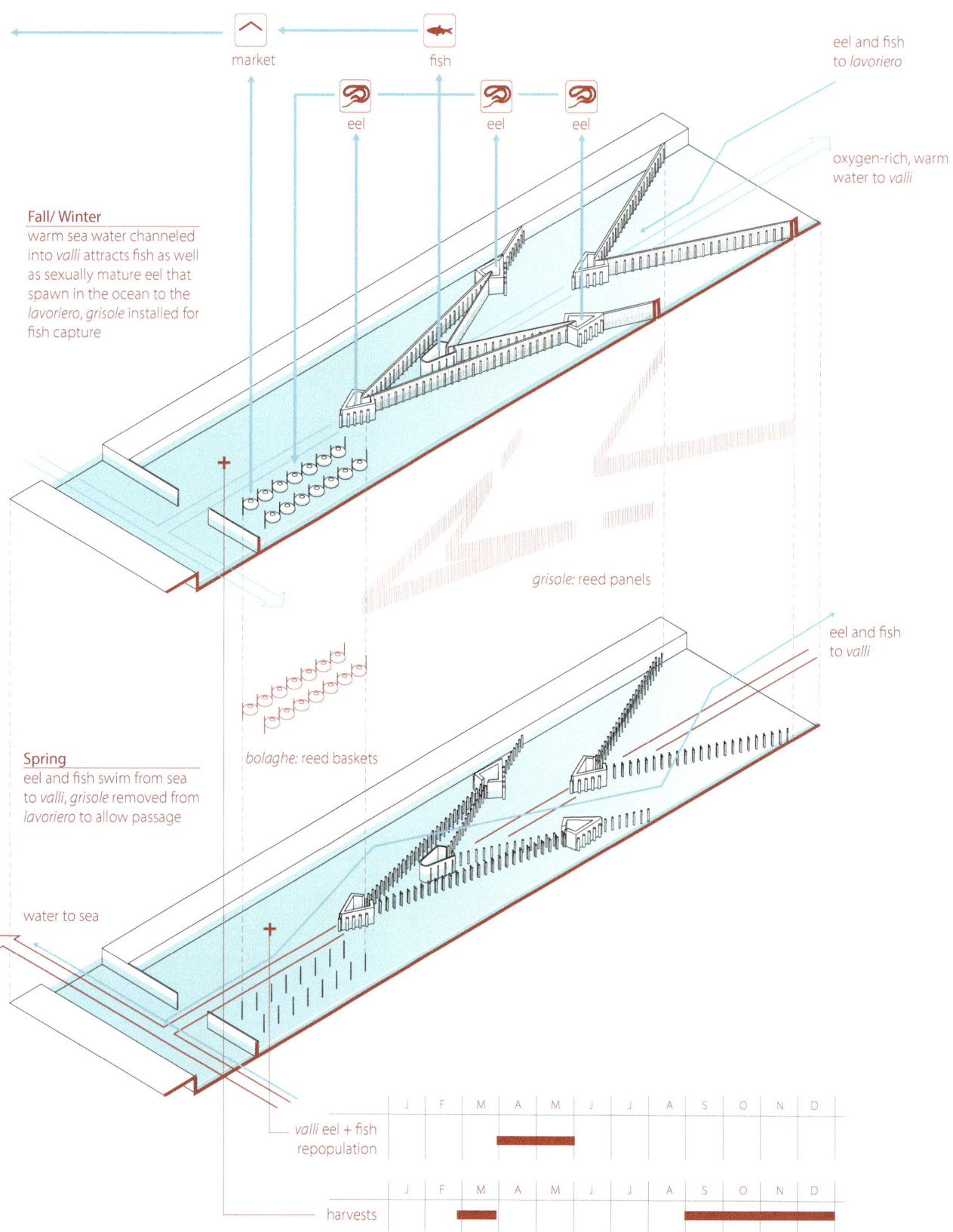

market

fish

eel and fish
to *lavoriero*

eel

eel

eel

oxygen-rich, warm
water to *valli*

Fall/ Winter
warm sea water channeled
into *valli* attracts fish as well
as sexually mature eel that
spawn in the ocean to the
lavoriero, *grisole* installed for
fish capture

grisole: reed panels

eel and fish
to *valli*

bolaghe: reed baskets

Spring
eel and fish swim from sea
to *valli*, *grisole* removed from
lavoriero to allow passage

water to sea

valli eel + fish
repopulation

	J	F	M	A	M	J	J	A	S	O	N	D

harvests

	J	F	M	A	M	J	J	A	S	O	N	D

09 *Landscape Systems and Strategies.* The canals embedded in the constructed dikes of the valli channel salt water deep into the brackish coastal lagoon, creating zones of hypersalinity that attract eels and fish into the lavorieri during fall and winter seasons. In the spring the lavorieri screens are removed to allow eels and fish into the valli.

10

11

10 Sluices at the Écluses Passo-Pedone used to control inflow of freshwater from the Po Reno into the former Valle Vacca.
11 The reconstructed lavoriero at the Stazione di Pesca Serilla is a stop on the daily boat tour of the Valli di Comacchio. This fishing station was located at the intersection of three valli; today the erosion of the dikes that enclosed these valli is evident.

01 Contemporary view of the piscina constructed at the "Grotto of Tiberius," a natural seaside grotto at the Villa of Tiberius. Reproductions of the statuary that once adorned these fish basins are displayed at the adjacent National Archaeological Museum of Sperlonga.

Piscina of Sperlonga, Italy

Landscape Type: piscina

Landscape Area: 0.5 ha

Aquaculture Yield: unknown

Aquaculture Type: extensive

Water Type: brackish

Piscinae are ancient Roman fishponds designed for both the cultivation and the human enjoyment of fish. They were constructed on the Italian peninsula during the later Roman Republic and the early Empire. Piscinae were often associated with the opulent *villae maritimae* (seaside villas) of the elite, and were constructed to maintain fish for consumption, as pets, and for market sale. James Higginbotham, a scholar of piscinae, writes, "Landscape and topography augmented by architecture were fundamental to villa design. Piscinae helped broaden the definition of landscape to include not only the seashore but also the sea itself."[1] Fifty-six coastal and upland piscinae have been uncovered and documented in central Italy.[2]

Seaside piscinae range in size from several hectares to the most modest garden ponds. The largest piscinae required considerable resources and technical expertise for construction. Basins were hewn into rock that provided a stable foundation for hydraulic concrete construction. The means to channel both freshwater and sea water into the basins, as well as the provision within the ponds of spaces of refuge for fish, were significant design elements. The piscinae were typically stocked with eel, sea bass, and mullet that had been harvested locally from the sea. Piscinae were also likely passively populated by migratory fish that would enter seeking coastal lagoon habitat.

The extant piscina at the imperial Villa of Tiberius, a summer residence for the Emperor Tiberius, is set in and around the "Grotto of Tiberius," a natural seaside grotto. This piscina is characterized by a large circular basin adjoined by a smaller rectangular basin, and is thought to have been created primarily to raise eels.[3] The piscina is a brackish environment, as it mixes together water from three freshwater springs issuing from fissures in the grotto, and sea water that enters via a constructed channel that connects the circular basin to the sea. These hydrologic inputs functioned to regulate oxygen levels, temperature, and fish waste in the basins. As the tide went out, the channel directed brackish water into the sea, which would entice migrating eels to travel up the conduit and enter the piscina. *Specus*, recesses typically built into the sidewalls of piscinae to offer fish a shaded refuge, are found in the rectangular basins of this piscina. A breakwater constructed several meters from the piscina protects it from wave action.

This aquaculture landscape offered vivid aesthetic experiences to privileged visitors. The piscina featured a *triclinium*, a covered dining area, dramatically positioned at the center of the fishpond. Visitors reclining on couches beneath the pergola at the triclinium would have had an ideal view to statuary groups within the grotto as they consumed food floated to them on small boats.[4]

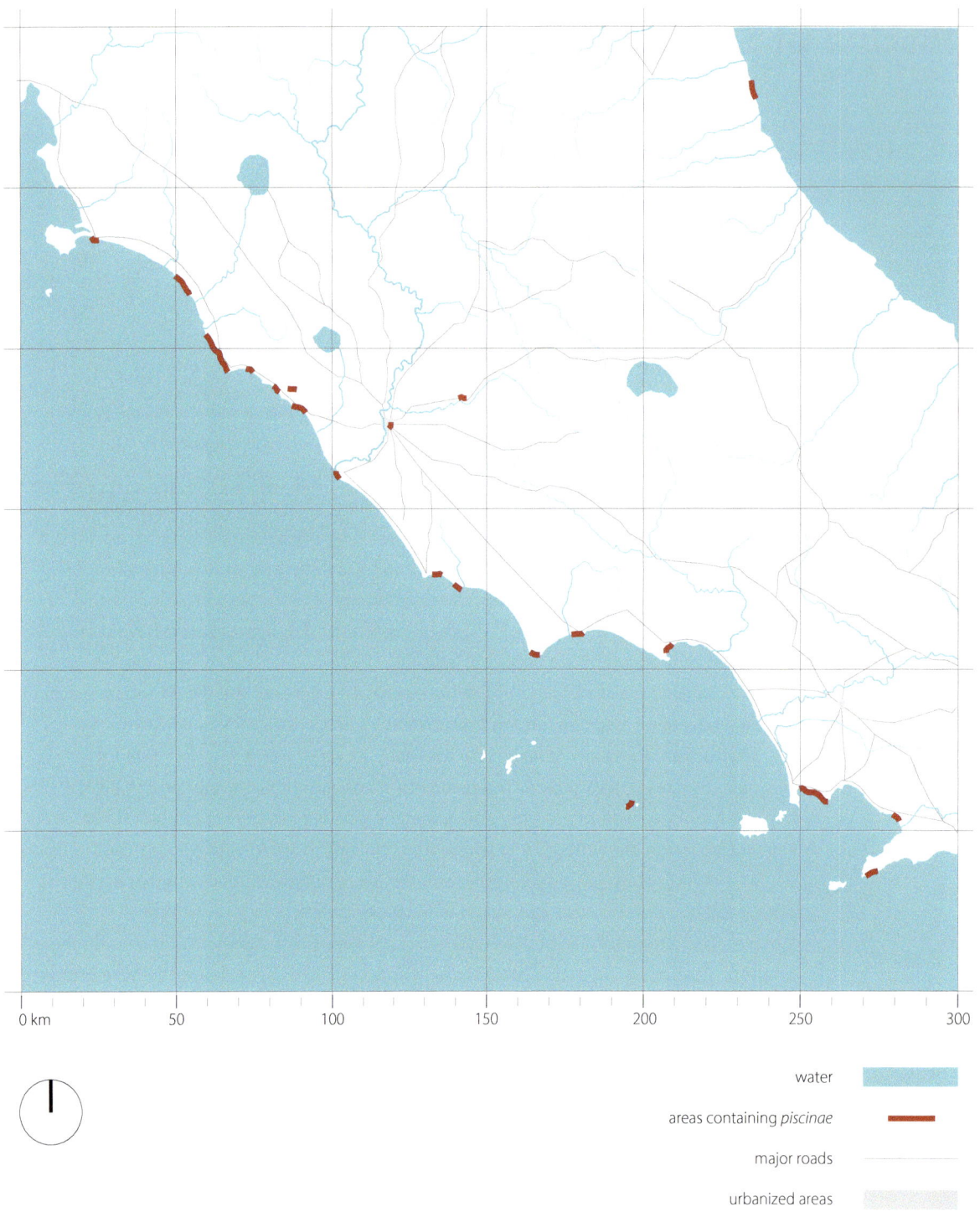

0 km   50   100   150   200   250   300

water

areas containing *piscinae*

major roads

urbanized areas

02   *Regional Diagram*. This map depicts the central Italian peninsula during the late Roman Republic/ early Roman Empire and locates areas where coastal and inland piscinae were constructed. This diagram reproduces a map by James Higginbotham, originally published in his book *Piscinae: Artificial Fishponds in Roman Italy*.

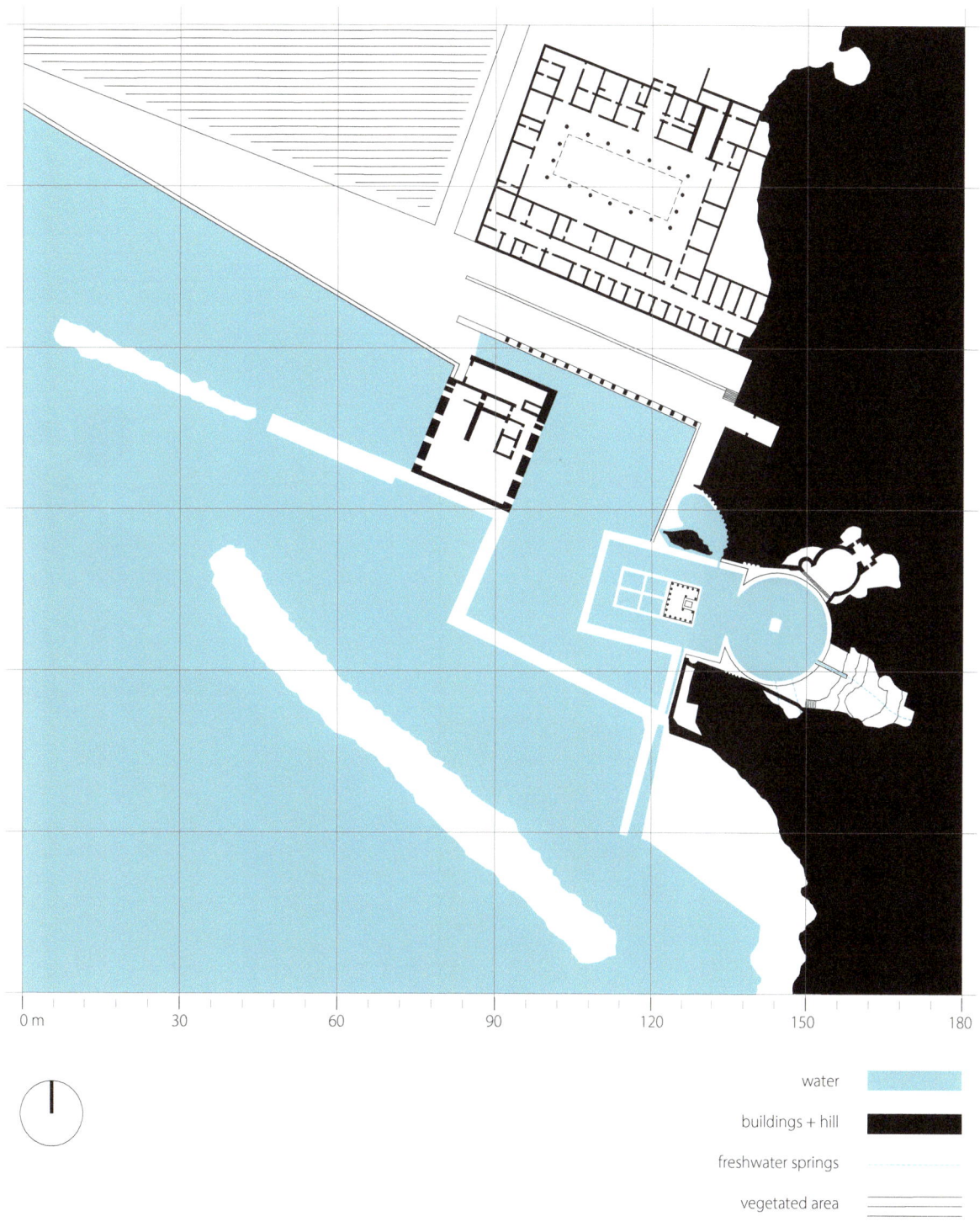

water

buildings + hill

freshwater springs

vegetated area

03 *Site Diagram.* The Villa of Tiberius featured a piscina that was constructed within and in front of a natural seaside grotto. This piscina is subdivided into five basins. This subdivision may have allowed for the separation of fish species to reduce predation, to gather fish selected for consumption, or to display fish kept as pets.

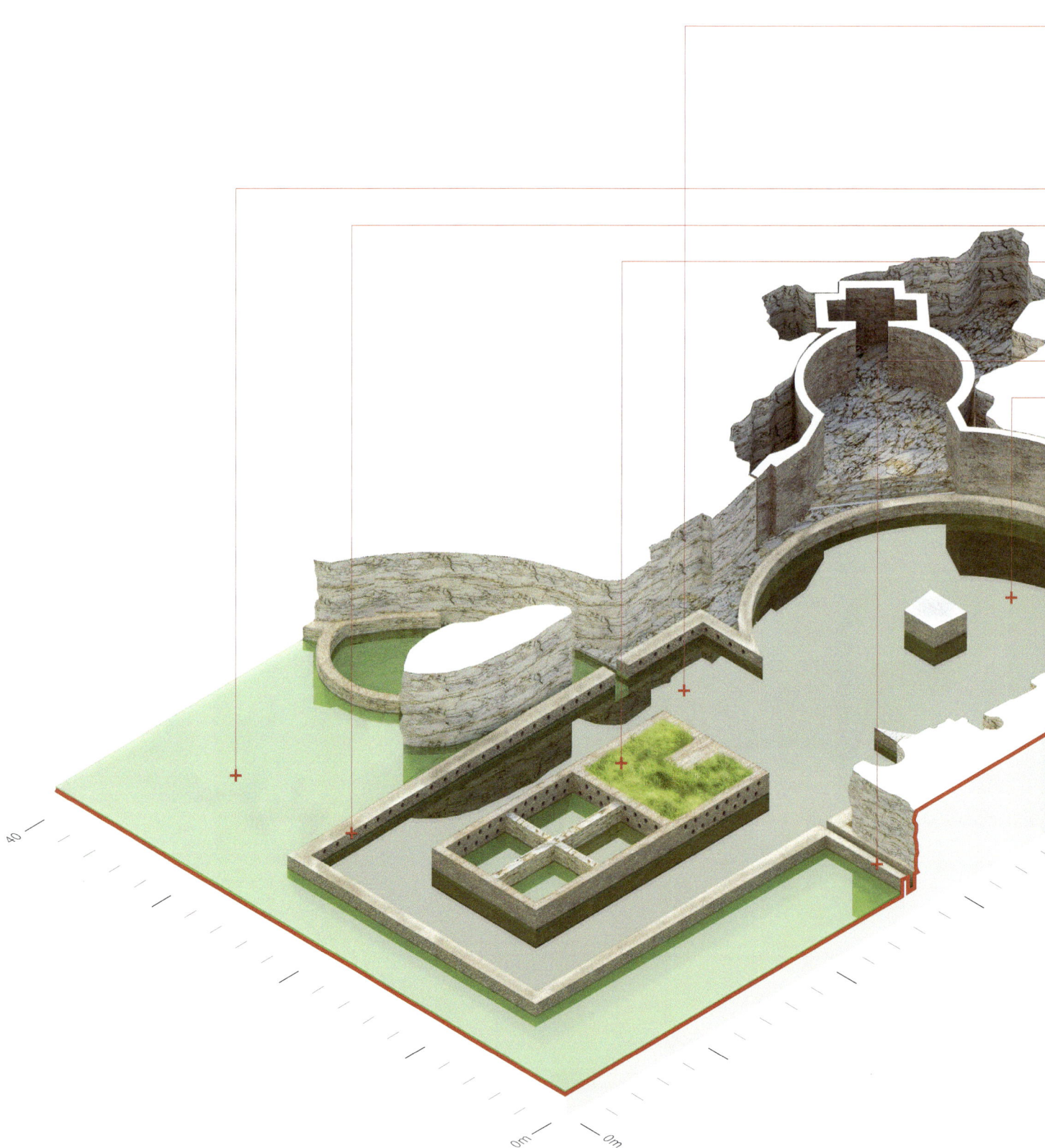

Cultivated Fauna

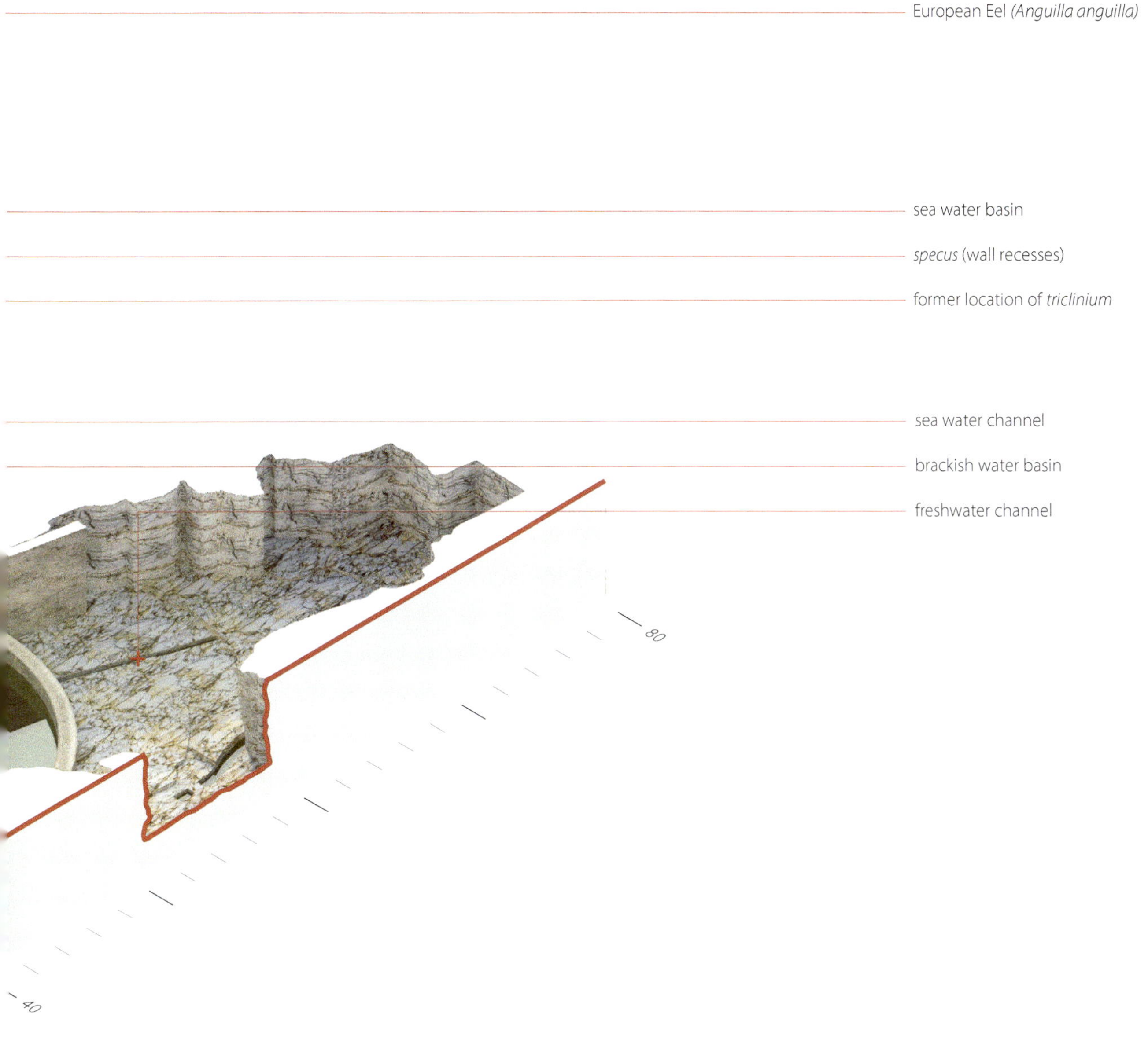

European Eel *(Anguilla anguilla)*

sea water basin

specus (wall recesses)

former location of *triclinium*

sea water channel

brackish water basin

freshwater channel

80

40

04 *Transect.* The basin sidewalls of the piscina were constructed of hydraulic concrete and faced in *opus reticulatum (Roman brickwork).* The basins of the piscina held water at a depth of 1.5 m. This transect depicts the contemporary condition of the piscina, but 2,000 years ago the piscina featured five statuary groups and a triclinium, a small roofed pavilion for dining.

05 Interior of natural grotto with view to the triclinium. The circular basin is approximately 21 m wide.
06 View to the sea water channel and breakwater.

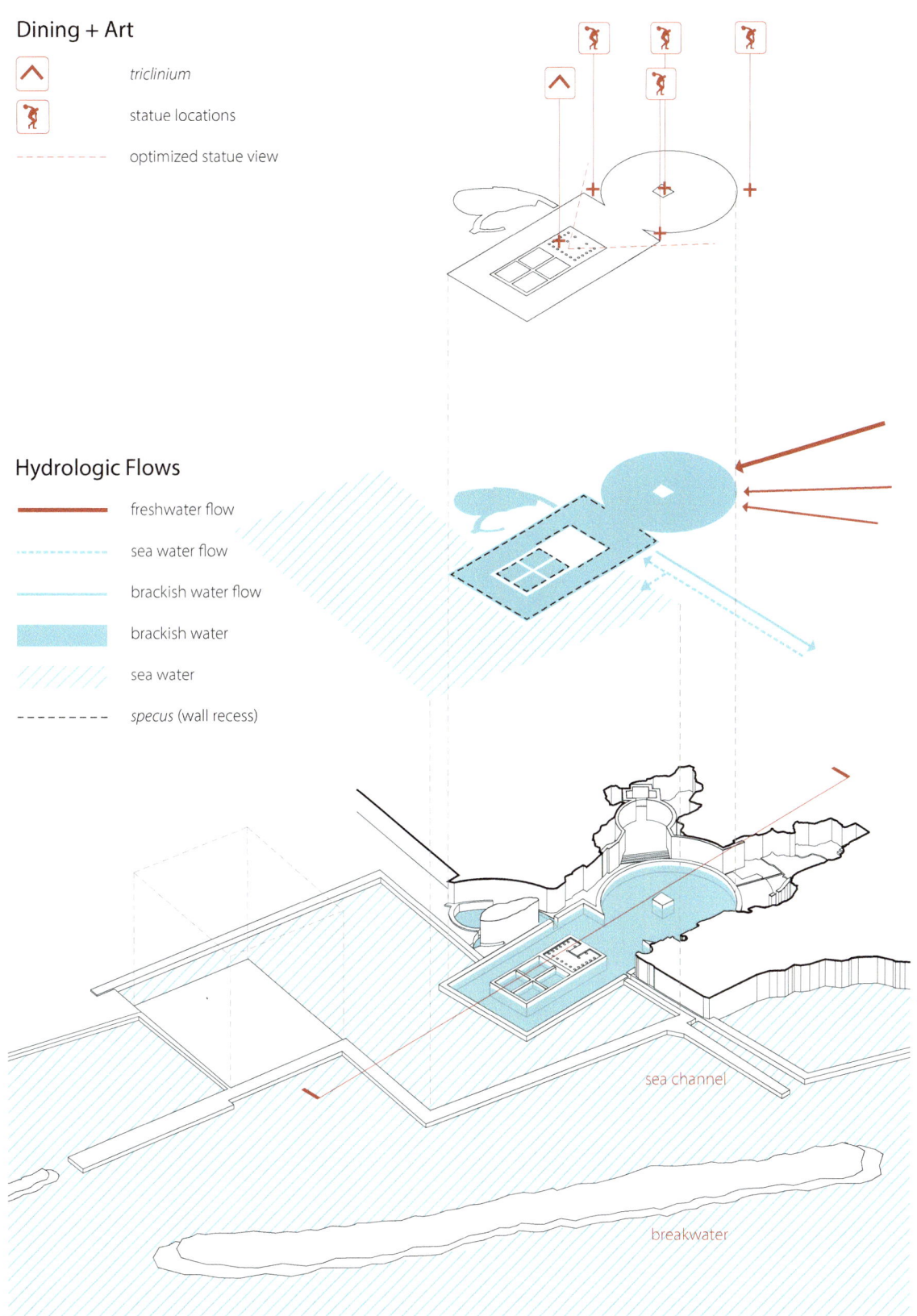

Dining + Art

⌃	*triclinium*
🤾	statue locations
- - - - - -	optimized statue view

Hydrologic Flows

————	freshwater flow
- - - - - -	sea water flow
————	brackish water flow
▓▓	brackish water
⁄⁄⁄⁄	sea water
- - - - - -	*specus* (wall recess)

sea channel

breakwater

07 *Landscape Systems and Strategies.* At the piscina at the Grotto of Tiberius a channel directed sea water into the basins, where it would mix with freshwater emanating from adjacent springs. A sluice at the channel was used to manage the quantity of salt water entering the basins, allowing for the salinity of the water to be optimized.

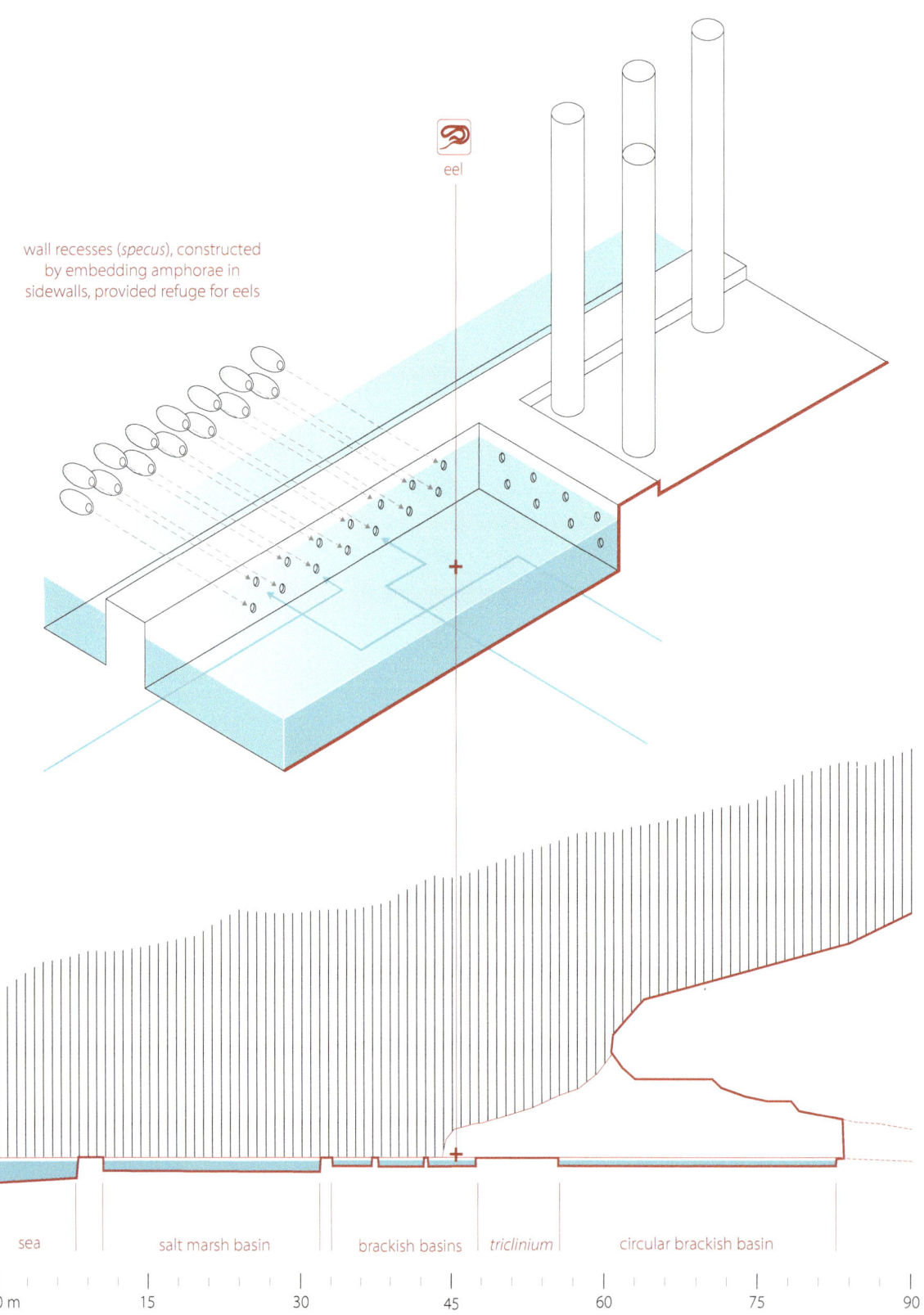

wall recesses (*specus*), constructed by embedding amphorae in sidewalls, provided refuge for eels

eel

sea   salt marsh basin   brackish basins | *triclinium*   circular brackish basin

0 m   15   30   45   60   75   90

08 *Landscape Systems and Strategies.* Eels find refuge in the specus (wall recesses) of the brackish water basins of the piscina. The specus were constructed by embedding clay amphora into walls constructed of hydraulic concrete. The relatively low freeboard at this piscina suggests that eels, which are not known to jump to escape confinement, were kept.

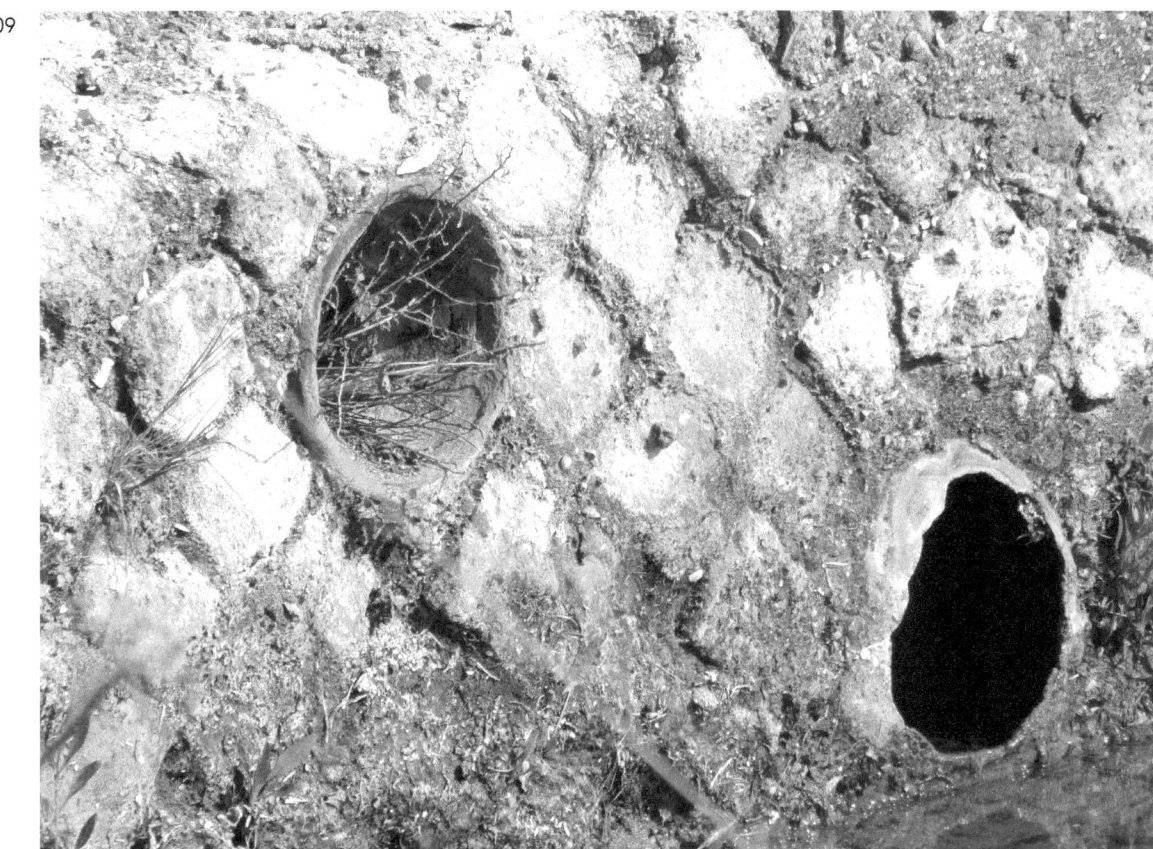

09 Detail of specus and opus reticulatum at sidewall of a brackish water basin adjacent to the triclinium.
10 View to circular brackish water basin with central statue pedestal.

01 Paul Langlois, detail from *Vue général de l'etablissement de pisciculture de Huningue*, 1888. Engraving of the hatchery landscape at Huningue thirty-six years after its construction. The constructed freshwater ponds depicted here featured treed islands and were used to raise carp.

Piscifactoire of Huningue, France

Landscape Type:	hatchery landscape
Landscape Area:	40 ha (piscifactoire, 19th C)
	900 ha (PCA, 2019)
Aquaculture Yield:	6-8,000,000 eggs/yr (1850s)
Aquaculture Type:	N/A
Water Type:	fresh

The first modern hatchery recorded in Western history was established near the city of Huningue in the Alsace region of France in 1852.[1] The hatchery was a state-sponsored, 40 ha aquaculture landscape that featured technologies and practices still utilized in contemporary hatchery facilities. The hatchery was designed for harvesting and distributing massive volumes of trout and salmon eggs to restock rivers blighted by industrialization. Eggs were shipped across France and Europe, and also to the Americas and Australia.[2] The hatchery came to be known as the "*piscifactoire*," or fish factory.

The scientist and academic, Victor Coste, together with two engineers, developed the sophisticated plan for the piscifactoire. Sited in the floodplain of the Rhine River, construction of the piscifactoire involved significant hydrologic manipulation of the landscape including rerouting river and canal water to outdoor fishponds, piping natural spring water for egg harvesting via sub-surface aqueducts, and managing wastewater in lagoons.[3] Coste's minimalist vision for the piscifactoire—a sub-divided and extended stream "covered with a glass roof like a greenhouse, admitting the light, and formed of movable panes turning around"—did not come to pass.[4] However, aquaculture historian Colin Nash writes, "The hatchery was far from utilitarian. The main building was constructed with observation galleries to accommodate the many distinguished visitors who flocked to see the phenomenon of fish culture. The facility was so impressive that many visitors to Huningue immediately returned home and replicated its basic concepts. . ."[5] Along with the viewing galleries which allowed delegations of scientists and curious citizens to observe egg extraction processes within the buildings, the hatchery landscape was an aesthetic experience of aquaculture designed for a public whose appetite for spectacle was being whetted by the advent of zoological gardens, aquaria, and international exhibitions. Walking trails complete with educational signage enabled visitors to stroll among the concentric arcs of nursery ponds, organic-shaped islands and basins for carp, and wooded experimental fishponds of this unique aquaculture landscape.[6]

In 1982, the piscifactoire and its alluvial forest ecosystem became part of the Réserve Naturelle de la Petite Camargue Alsacienne (PCA), a 900 ha multinational nature reserve. The PCA incorporates conservation practices and recreational activities; fish egg production, migratory bird ringing and Highland cattle rearing for meadow management all share space with trails and bird blinds. Today the nineteenth-century fishponds, once stocked with catfish, perch and carp, provide multiple ecosystem services, including flood water retention, provision of wetland habitat, nutrient filtration, and aquifer recharge.

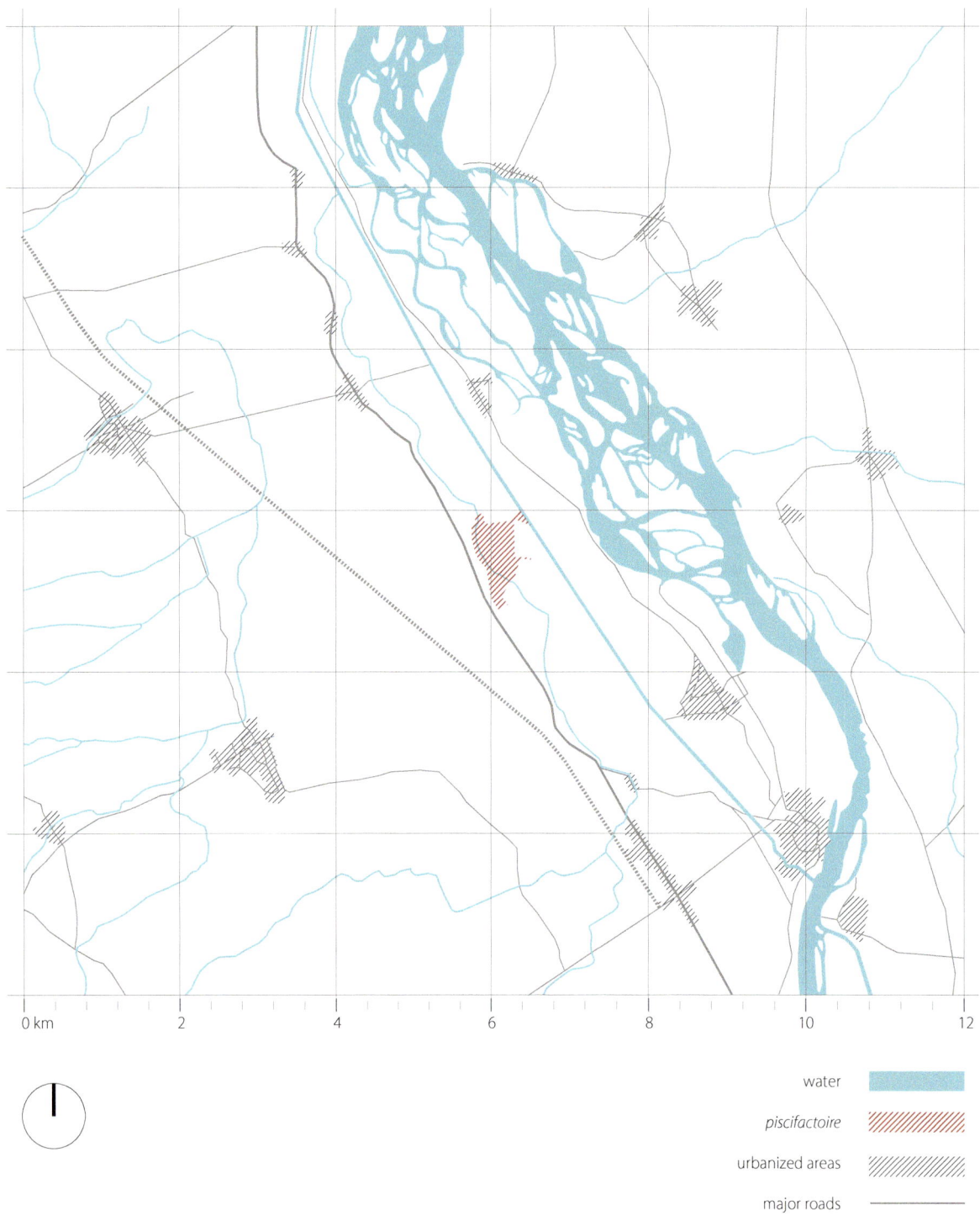

0 km　　2　　4　　6　　8　　10　　12

water

piscifactoire

urbanized areas

major roads

02 *Regional Diagram.* This diagram depicts the context of the piscifactoire in 1852, subsequent to the construction of the Huningue Canal but prior to the canalization of the meandering Rhine River. Industrialization and transformation of rivers across France decimated fish populations and reinforced the need for fish egg production and fish stocking.

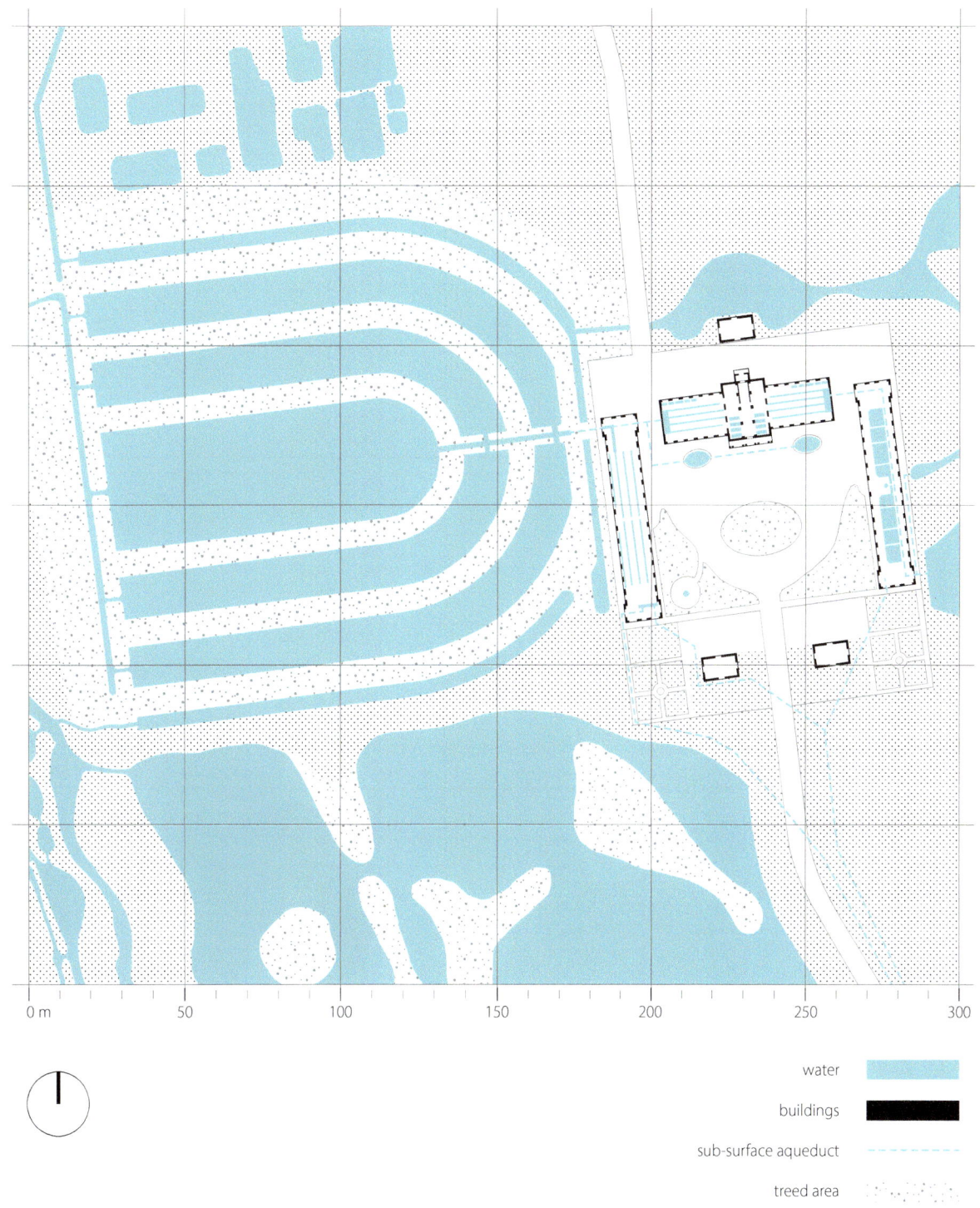

0 m   50   100   150   200   250   300

water

buildings

sub-surface aqueduct

treed area

03 *Site Diagram*. The piscifactoire was a 40 ha landscape constructed for fish production and experimentation and served by a sophisticated hydrological system. Nursery ponds are shaped in concentric arcs and shaded with planted trees. The ornate hatchery buildings frame a formal courtyard and gardens.

Cultivated Fauna

Atlantic Salmon *(Salmo salar)*
Danube Salmon *(Hucho hucho)*
Brown Trout *(Salmo trutta)*
char *(Salvelinus spp.)*
Lake Herring *(Coregonus artedi)*

Wels Catfish *(Silurus glanis)*
Tench *(Tinca tinca)*
pike *(Esox spp.)*
perch *(Perca spp.)*
carp *(Cyprinus spp.)*

Cultivated Flora

oak *(Quercus spp.)*
black poplar *(Populus nigra)*

wastewater drainage canal

fish egg harvesting troughs

water supply trench

fingerling grow-out basins

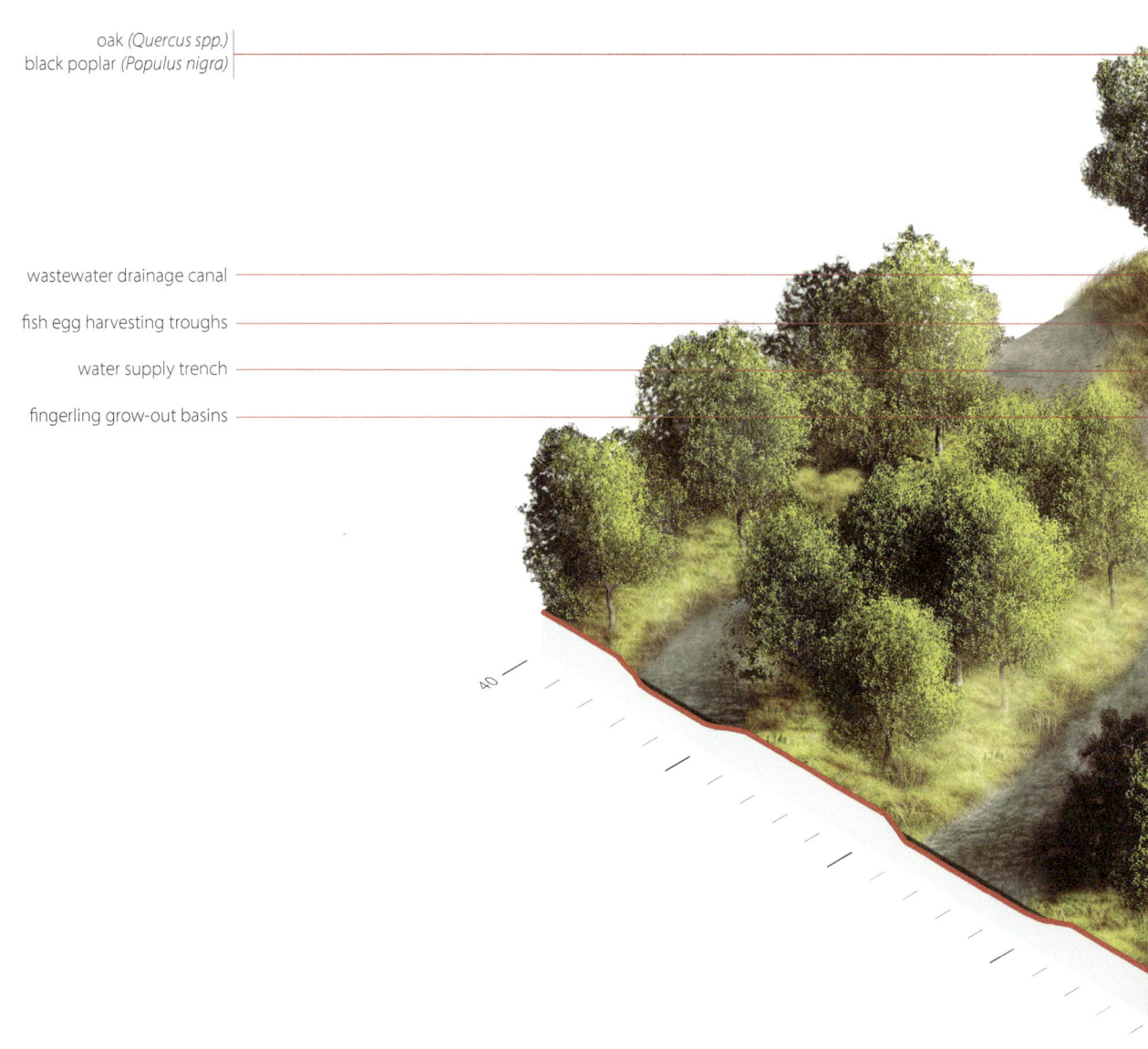

40

0m

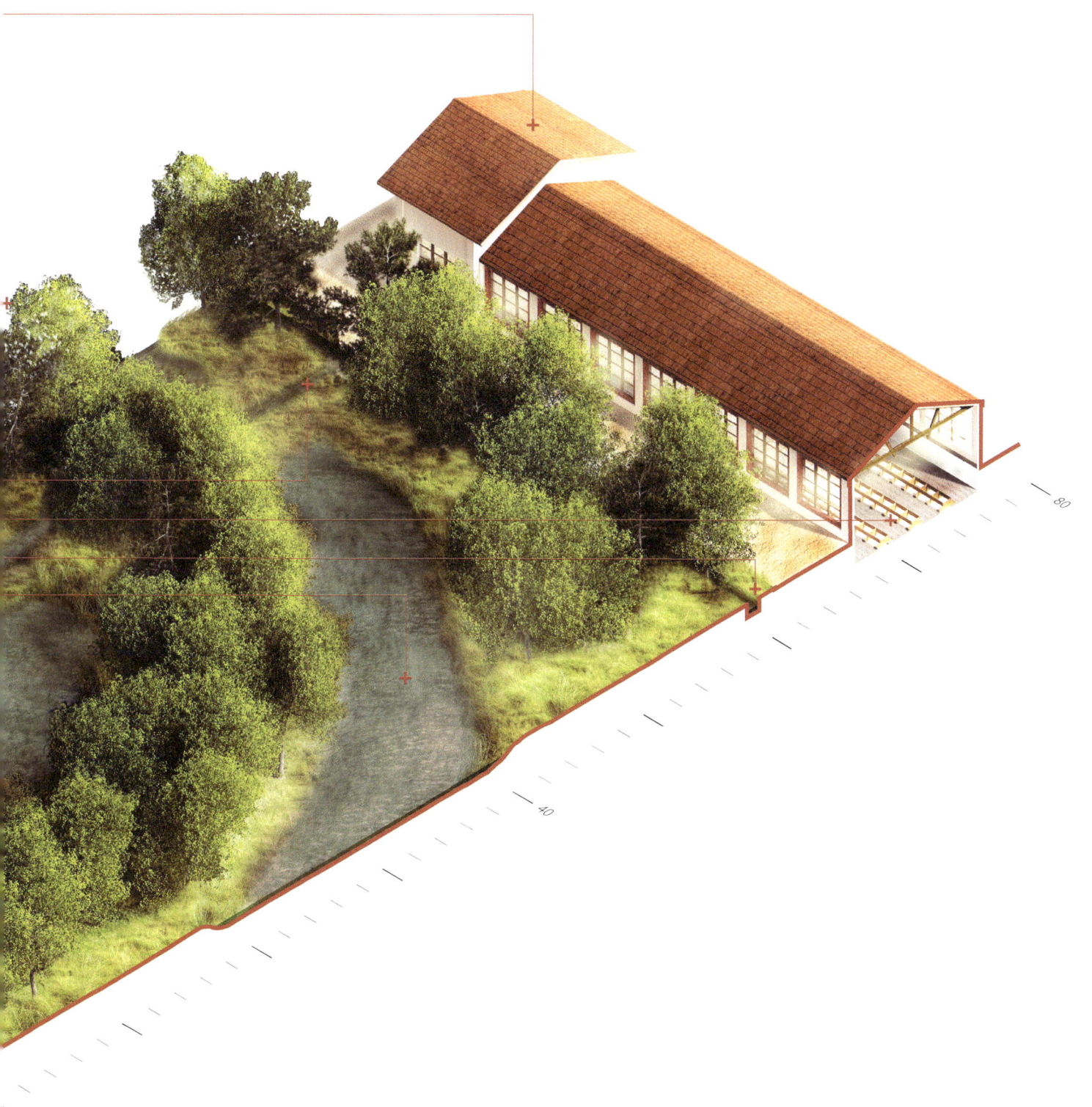

04 *Transect.* The fingerling cultivation ponds depicted in this transect were supplied in part with water discharged from the adjacent hatchery buildings. Water was supplied to the egg harvesting troughs in the buildings via sub-surface aqueducts.

05 Mason Jackson, detail from *The Fish Nursuries at Huningue, France,* 1861. An etching of the view to the main buildings and courtyard at the piscifactoire.
06 Contemporary tourists observe wildlife in the ponds directly in front of the historical buildings of the piscifactoire.

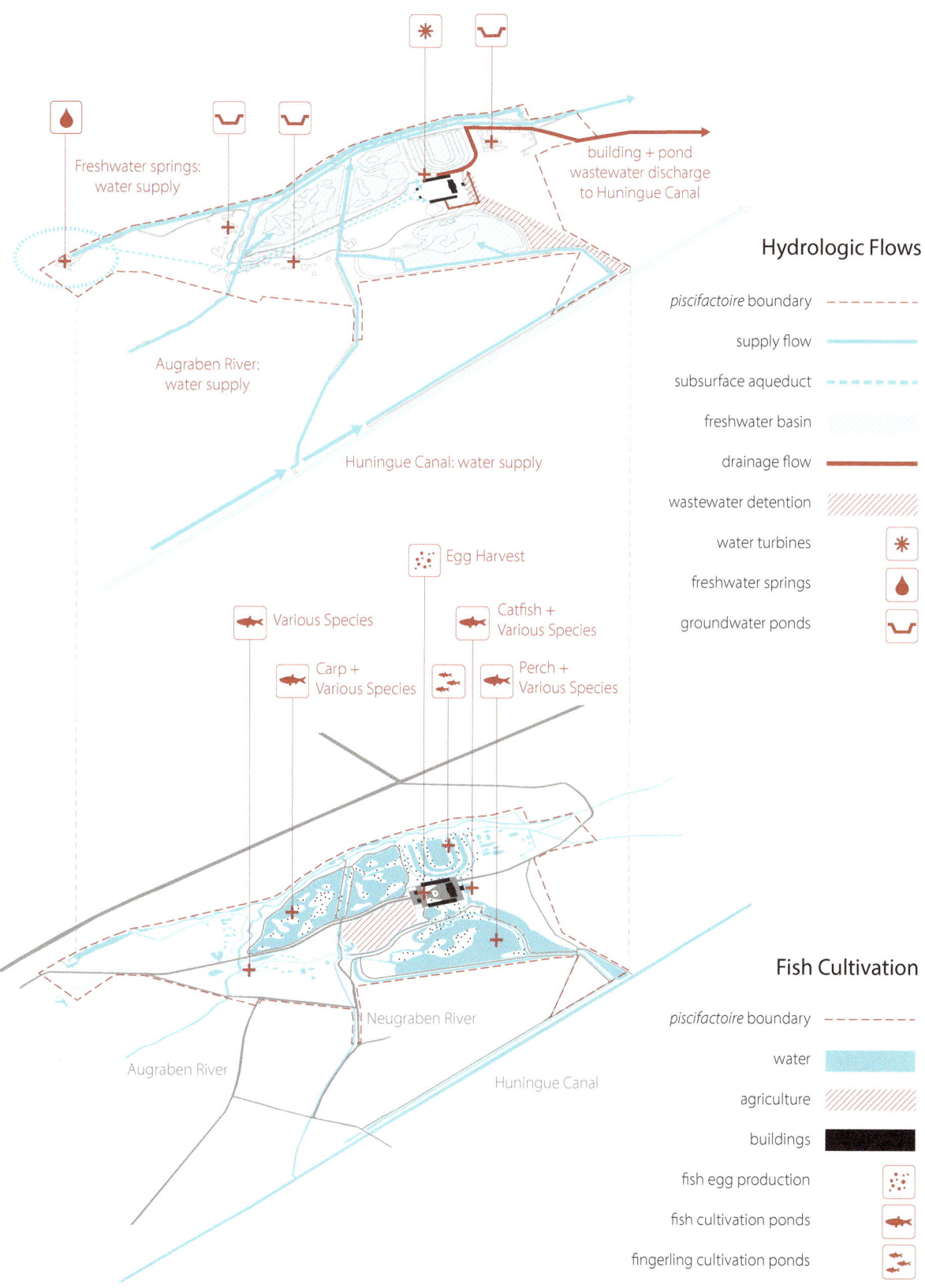

Hydrologic Flows

piscifactoire boundary	----------
supply flow	——
subsurface aqueduct	- - - - - -
freshwater basin	//////
drainage flow	——
wastewater detention	//////
water turbines	✳
freshwater springs	🜄
groundwater ponds	⌣

Freshwater springs: water supply

Augraben River: water supply

Huningue Canal: water supply

building + pond wastewater discharge to Huningue Canal

Egg Harvest

Various Species

Carp + Various Species

Catfish + Various Species

Perch + Various Species

Neugraben River

Augraben River

Huningue Canal

Fish Cultivation

piscifactoire boundary	----------
water	▓
agriculture	//////
buildings	■
fish egg production	⋮
fish cultivation ponds	🐟
fingerling cultivation ponds	🐟🐟

07 *Landscape Systems and Strategies*. The ponds and buildings of the piscifactoire are supplied with water from three or more sources. Water from freshwater springs is pumped and piped to the buildings for use in egg-harvesting processes, while water from local rivers and the Huningue Canal supplies exterior ponds where various species of fish are reared.

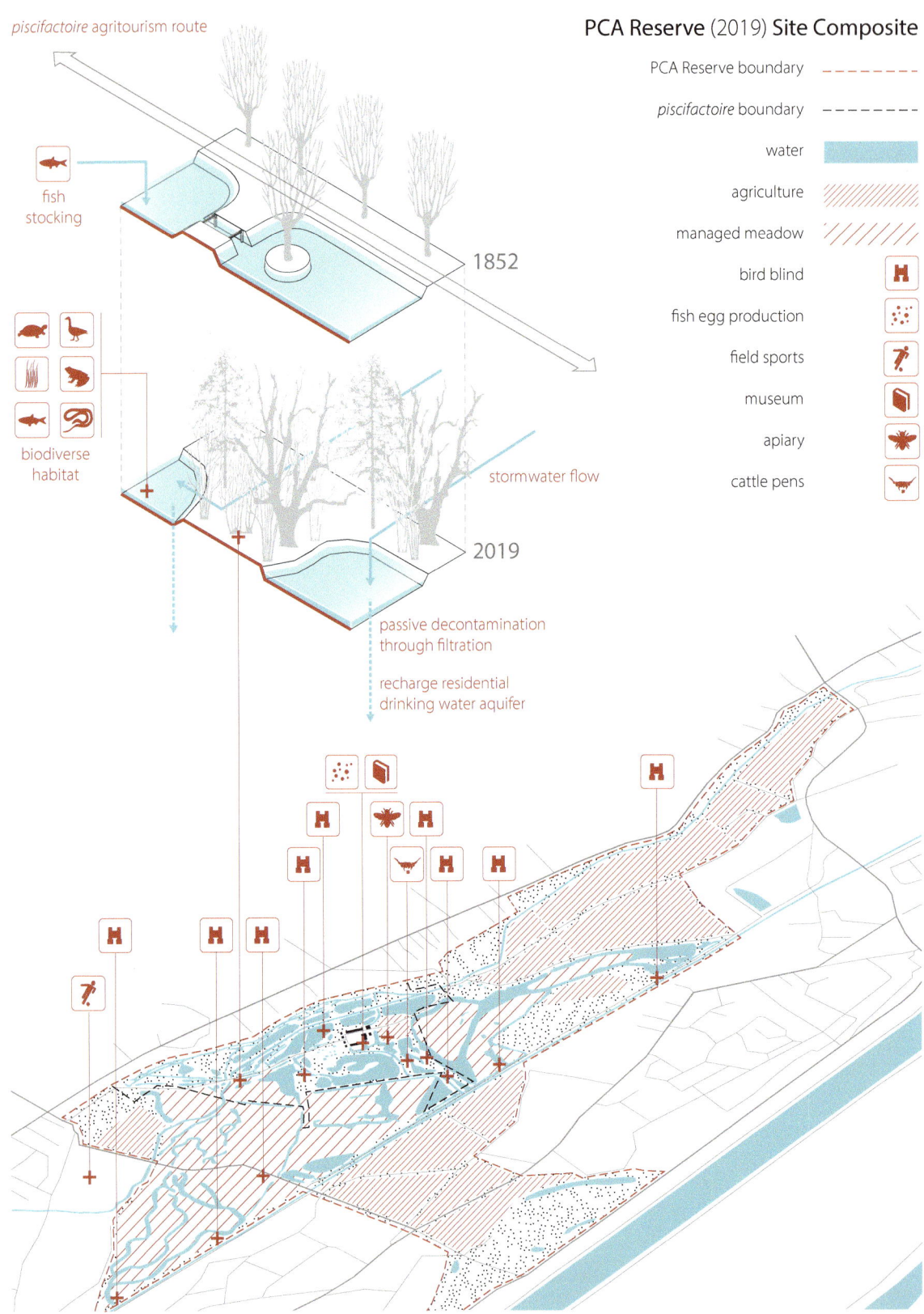

PCA Reserve (2019) Site Composite

piscifactoire agritourism route

fish stocking

biodiverse habitat

1852

2019

stormwater flow

passive decontamination through filtration

recharge residential drinking water aquifer

PCA Reserve boundary

piscifactoire boundary

water

agriculture

managed meadow

bird blind

fish egg production

field sports

museum

apiary

cattle pens

08 *Landscape Systems and Strategies.* Today the piscifactoire is part of a 900 ha multinational nature reserve. This managed alluvial forest ecosystem provides ecological services—drinking water aquifer recharge, passive decontamination, core habitat—and supports ecotourism and active recreation along the urbanized and industrialized Rhine River.

09

10

09 Adolphe Braun, *Etablissement de pisciculture de Huningue No. 5*, 1862. View to the main building of the piscifactoire from the courtyard. The building included fish egg harvesting troughs, public viewing galleries, and hatchery worker housing.
10 Tourists utilize a series of boardwalks to access the interior of the nature reserve of which the piscifactoire is a part.

01 Osvaldo Tofani, detail from *L'aquarium du Trocadéro, L'Exposition Universelle*, 1878. The subterranean gallery of the grotto aquarium featured water-filled basins that housed a wide range of fish species, as well as a dramatic interior water cascade that opened to the gardens above.

Grotto Aquarium in Paris, France

Landscape Type: grotto aquarium

Landscape Area: 0.32 ha

Aquaculture Yield: ~150,000 eggs/yr | ~60,000 fry/yr (19th C)

Aquaculture Type: intensive

Water Type: fresh

The Aquarium du Trocadéro, an innovative and immersive public landscape, was built to display fish at the *Exposition Universelle* of 1878 in Paris and later served as a center for aquaculture in France. The aquarium was an element within the larger exposition gardens that were designed by J.C.A. Alphand and constructed adjacent to the Palais du Chaillot in the Trocadéro area of the sixteenth-century *arrondissement*.

The grotto aquarium was constructed within a pre-existing quarry pit. Upon descending open-air stairways and entering the subterranean gallery, visitors were immersed in a seemingly natural cave that featured rockwork and imitation stalactites constructed using *stuc ciment* (stucco cement). Twenty-three open-air, 4 m deep basins lined the curving walls of an elliptical promenade, and each basin featured windows to allow observation of dozens of fish species.[1] The grotto aquarium was roofed with an ornamental garden, complete with walking paths, exotic plantings, ornate kiosks, and curving streams and ponds with stone sidewalls and weirs. These water features were, in fact, the tops of the basins that held fish in the gallery below. Water flowed through this interconnected basin system, and was oxygenated by cascading over the weirs.

Thomas Ferguson, a United States Commissioner to the 1878 fair, described the aesthetics of water, light, and fish at this multi-level landscape. He noted that from the aquarium garden, the fish basins had the "pleasing appearance of a rivulet spanned by numerous rustic bridges and skirted by artistically planned and well-arranged walks," whereas in the grotto, which was softly illuminated by sunlight filtering through the basin windows, hundreds of fish were "flashing back the rays of the sun from their burnished sides."[2] The subterranean gallery was animated by another dramatic water feature—a cascade of water splashed five meters down from the gardens above into a pond at the one end of the gallery. Many of the aesthetic aspects of the gallery and garden—stuc ciment rockwork, *faux bois* (false wood) bridge railings, an internal cascade—were first utilized at the Parc des Buttes Chaumont, constructed in 1867 and designed by Alphand.[3]

In the years following the exposition, the grotto aquarium expanded its function and became a center of aquaculture experimentation, education, and production in France. In the nineteenth century, the aquarium produced over one million large salmonoid fry for restocking rivers in France and established relations with over 140 emerging aquaculture and angling societies across the country. The aquarium offered technical aquaculture courses to an interested public and experimental fish reproduction practices were ongoing.[4] Renovations in the late twentieth century, which followed an extensive expansion in 1937, erased all trace of its sublime grotto aesthetic.

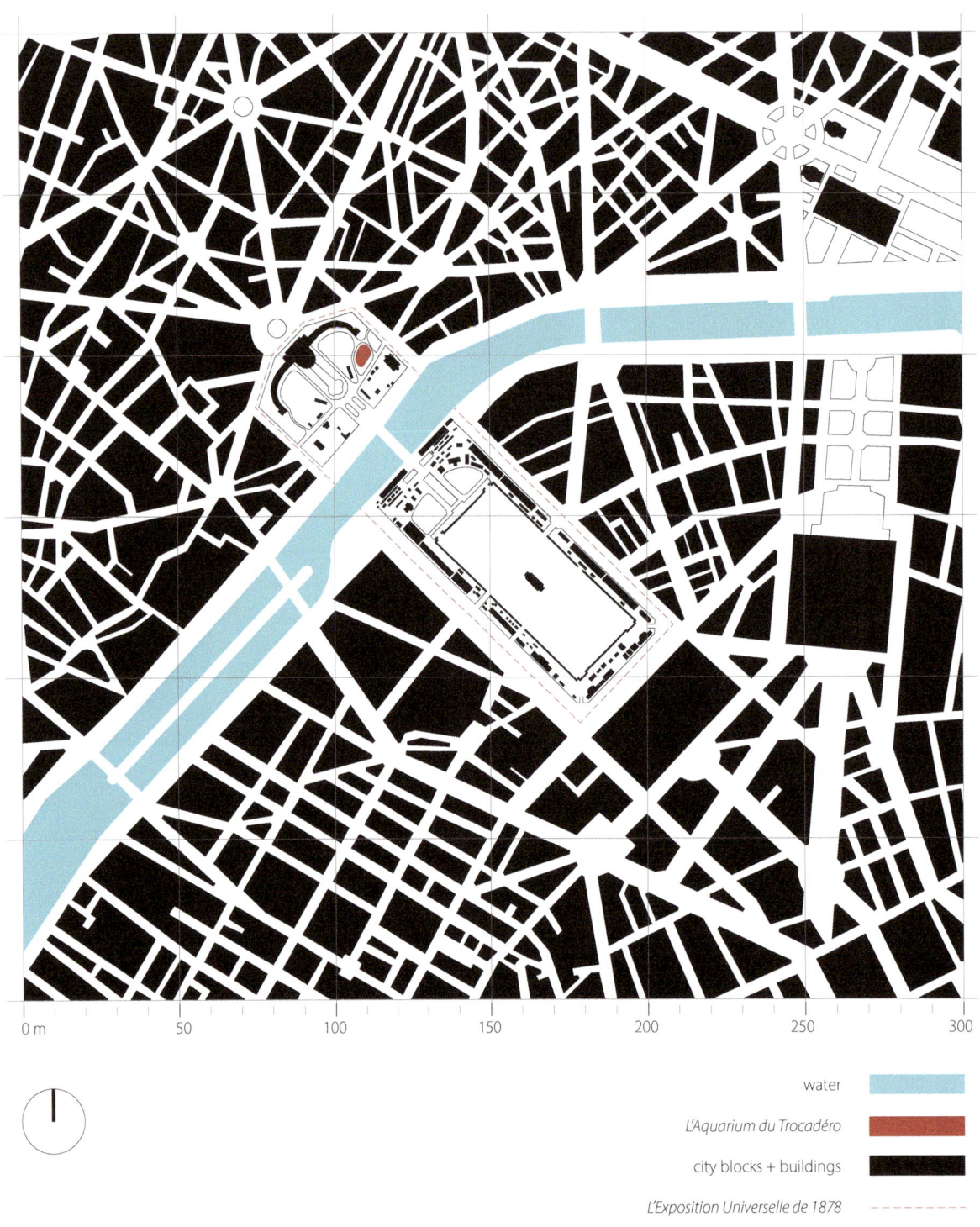

water	████
L'Aquarium du Trocadéro	████
city blocks + buildings	████
L'Exposition Universelle de 1878	---------

02 *Regional Diagram.* The Exposition Universelle of 1878, located at the Champ de Mars and the hill of Chaillot at Trocadéro, covered 27 ha and spanned the Seine River. The Aquarium du Trocadéro was constructed for the exposition in the gardens of the Palais du Trocadéro, which were designed by Jean-Charles Alphand.

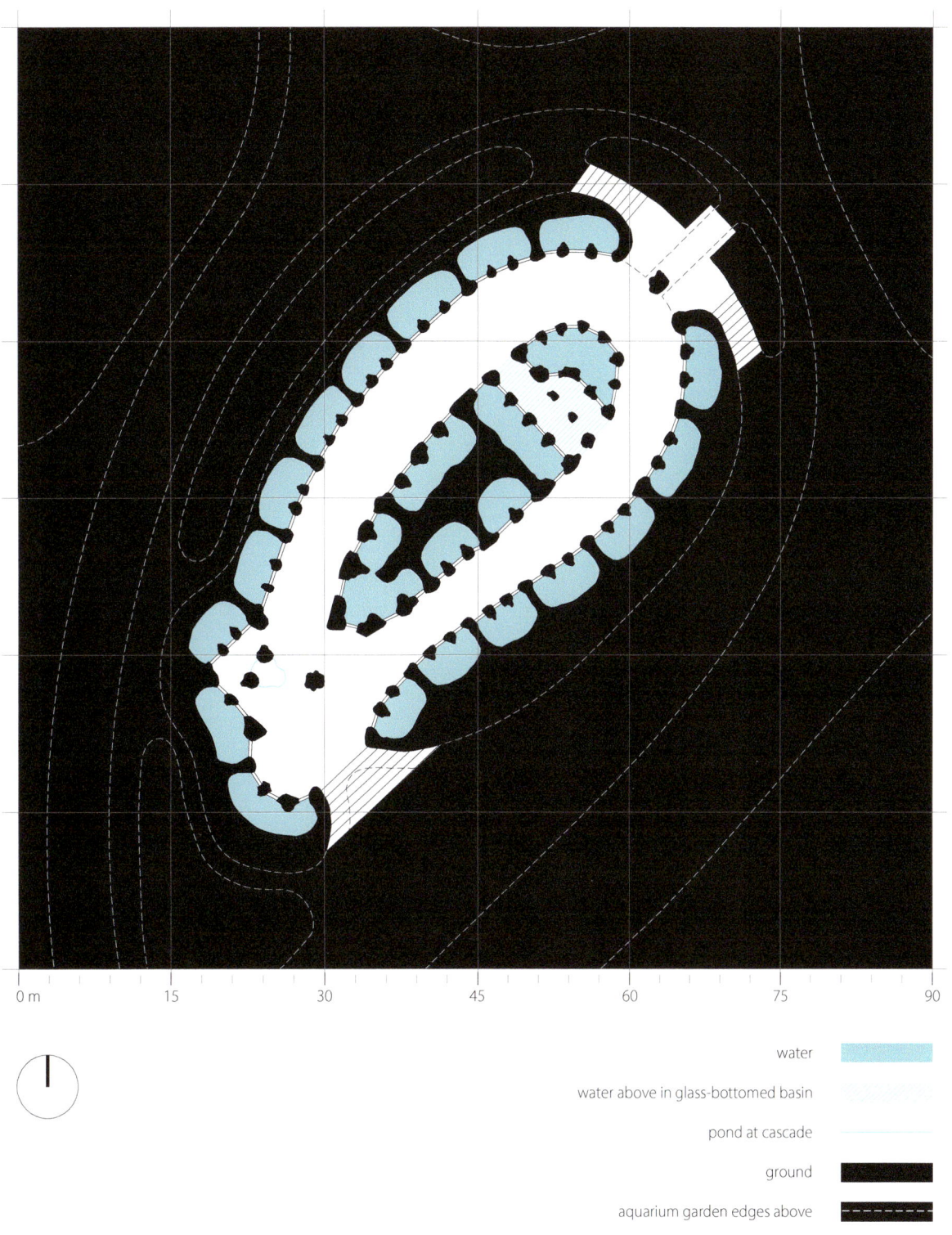

water
water above in glass-bottomed basin
pond at cascade
ground
aquarium garden edges above

0 m   15   30   45   60   75   90

03 *Site Diagram.* The subterranean gallery of the aquarium was constructed in a derelict quarry pit. The gallery was organized around an elliptical promenade and featured twenty-three individual basins. Visitors descended into the gallery from the garden above via three open-air stairways.

Cultivated Fauna

Danube Salmon *(Hucho hucho)*
Whitefish *(Coregonus lavaretus)*
Bavarian Char *(Salmo salvelinus)*
Rudd *(Scardinius erythrophthalmus)*
Chub *(Leuciscus cephalus)*
Brown Trout *(Salmo fario)*
Gudgeon *(Gobio gobio)*
Burbot *(Lota lota)*
Red-clawed Crab *(Perisesarma bidens)*
European Eel *(Anguilla anguilla)*
Common Bleak *(Alburnus lucidus)*
Sea Lamprey *(Peromyzon marinus)*
Common Barbel *(Barbus vulgaris)*
Common Bream *(Abramis brama)*
Common Nase *(Chondrostoma nasius)*
Wels Catfish *(Silurus glanis)*
Northern Pike *(Esox lucius)*
White Bream *(Abramis blicca)*
Ide *(Leuciscus idus)*
Tench *(Tinca tinca)*
Common Carp *(Cyprinus carpio)*

Cultivated Flora

oleaster *(Elaeagnus spp.)*
loquat *(Eriobotrya japonica)*
cypress *(Cupressus spp.)*
pondweed *(Elodea canadensis)*

stair to subterranean gallery

water over glass ceiling

fish basin observation window

gallery interior

road

weir between fish basins

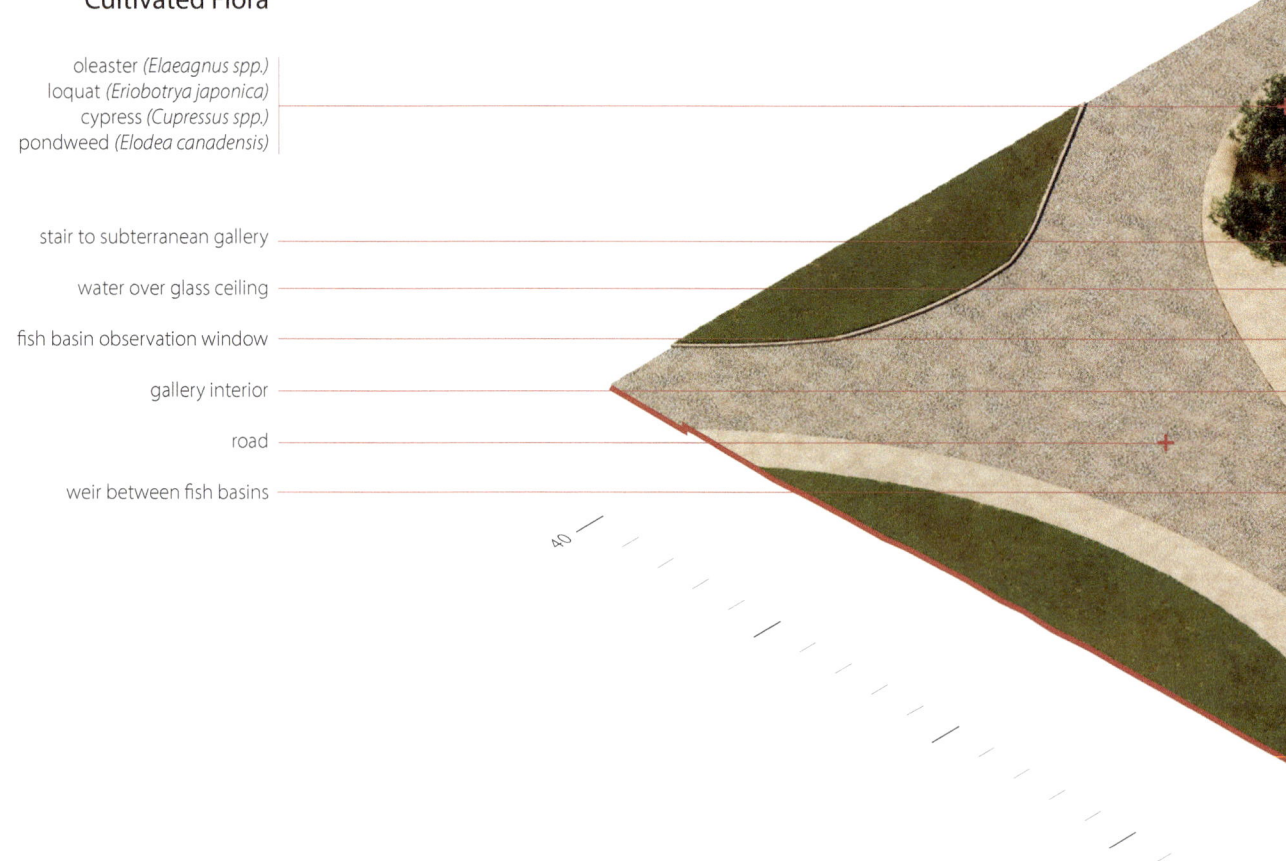

40

0m

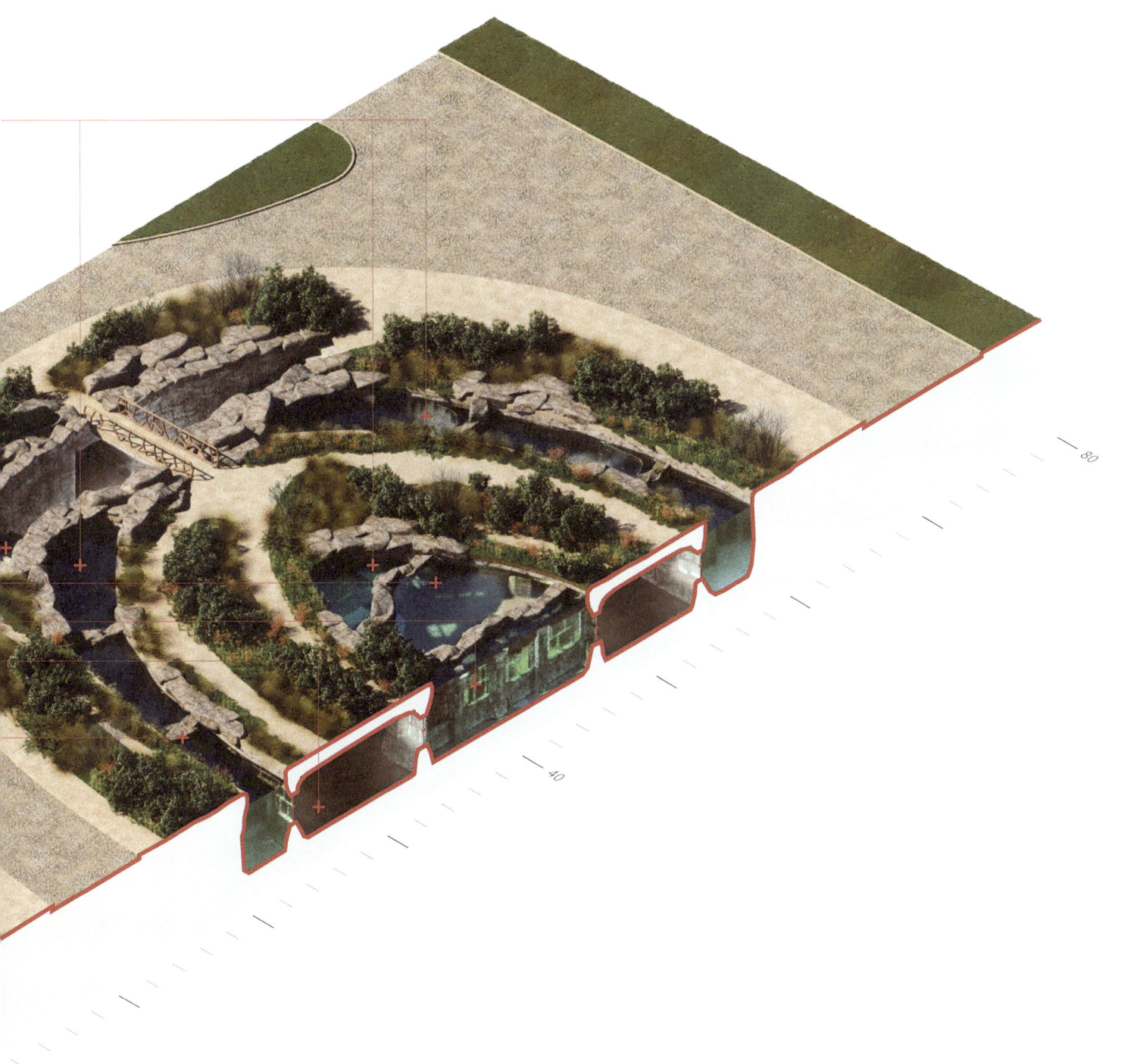

04 *Transect*. The Aquarium du Trocadéro was a multi-level landscape. The aquarium was roofed with a garden that featured walking paths, faux bois (false wood) bridges, and curving rivulets. The subterranean gallery was characterized by surfaces sculpted with stucco cement to impart a cave-like quality, and basin windows that glowed with sunlight.

05 Artist unknown, *Vue extérieur de l'aquarium du Trocadéro*, 1878. View to the faux bois bridge, central kiosk, and paths through the aquarium garden.
06 Agence de presse Meurisse, *Jardin du Trocadéro, vue de l'aquarium (intérieur)*, 1928. View to the subterranean gallery.

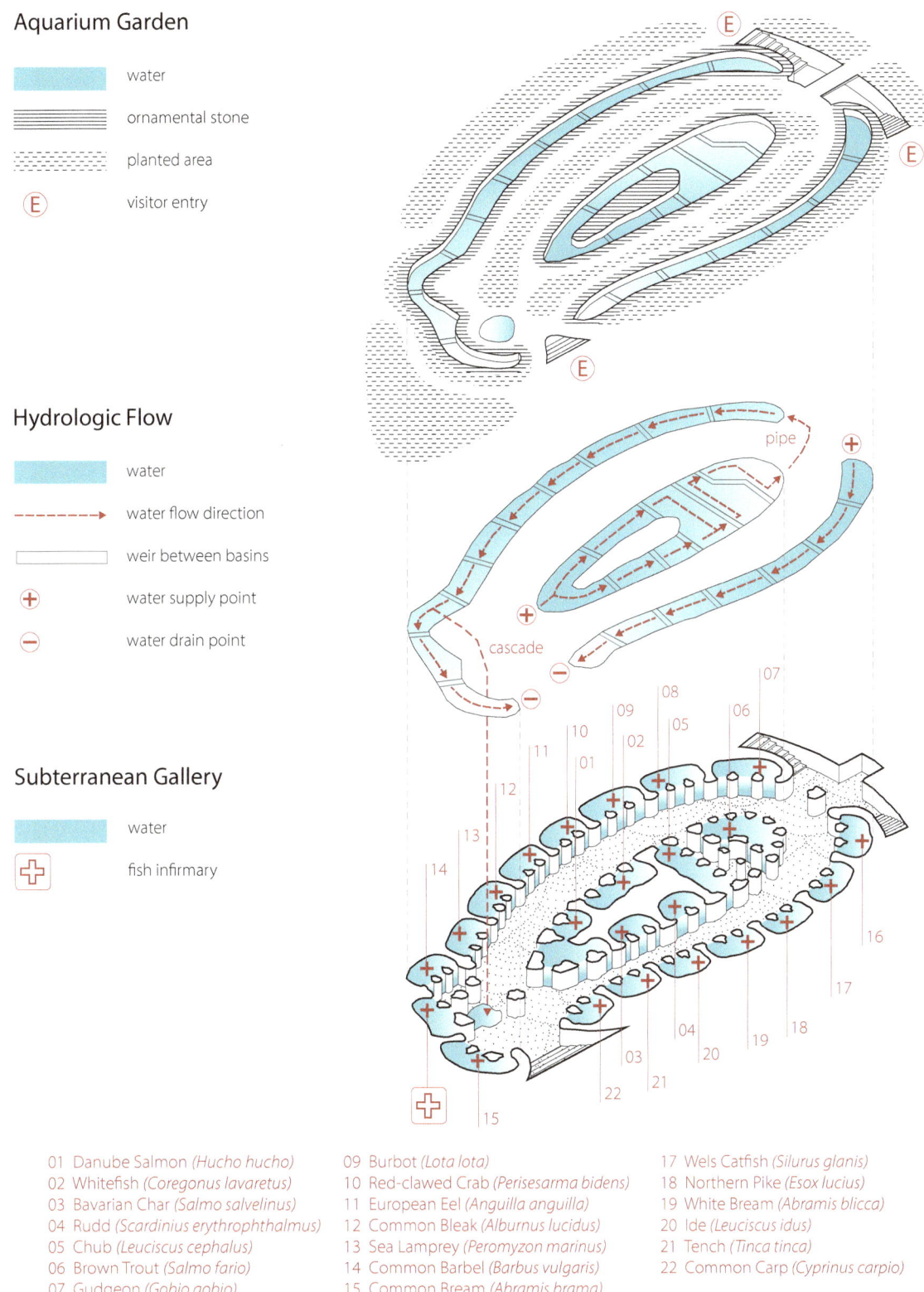

Aquarium Garden

▦ (water color)	water
▤ (stone pattern)	ornamental stone
▨ (dot pattern)	planted area
Ⓔ	visitor entry

Hydrologic Flow

▦ (water color)	water
------→	water flow direction
▭	weir between basins
⊕	water supply point
⊖	water drain point

Subterranean Gallery

▦ (water color)	water
✚	fish infirmary

01 Danube Salmon (*Hucho hucho*)
02 Whitefish (*Coregonus lavaretus*)
03 Bavarian Char (*Salmo salvelinus*)
04 Rudd (*Scardinius erythrophthalmus*)
05 Chub (*Leuciscus cephalus*)
06 Brown Trout (*Salmo fario*)
07 Gudgeon (*Gobio gobio*)
08 Ide (*Idus melanotus*)
09 Burbot (*Lota lota*)
10 Red-clawed Crab (*Perisesarma bidens*)
11 European Eel (*Anguilla anguilla*)
12 Common Bleak (*Alburnus lucidus*)
13 Sea Lamprey (*Peromyzon marinus*)
14 Common Barbel (*Barbus vulgaris*)
15 Common Bream (*Abramis brama*)
16 Common Nase (*Chondrostoma nasius*)
17 Wels Catfish (*Silurus glanis*)
18 Northern Pike (*Esox lucius*)
19 White Bream (*Abramis blicca*)
20 Ide (*Leuciscus idus*)
21 Tench (*Tinca tinca*)
22 Common Carp (*Cyprinus carpio*)

07 *Landscape Systems and Strategies.* Each of the twenty-three gallery basins held a different species of fish. At the garden level, water flowed over weirs between basins, and the gradually descending height of weirs in the system allowed water to flow from input to drain. Individual basins were grouped into long, curving rivulets through stone perimeters.

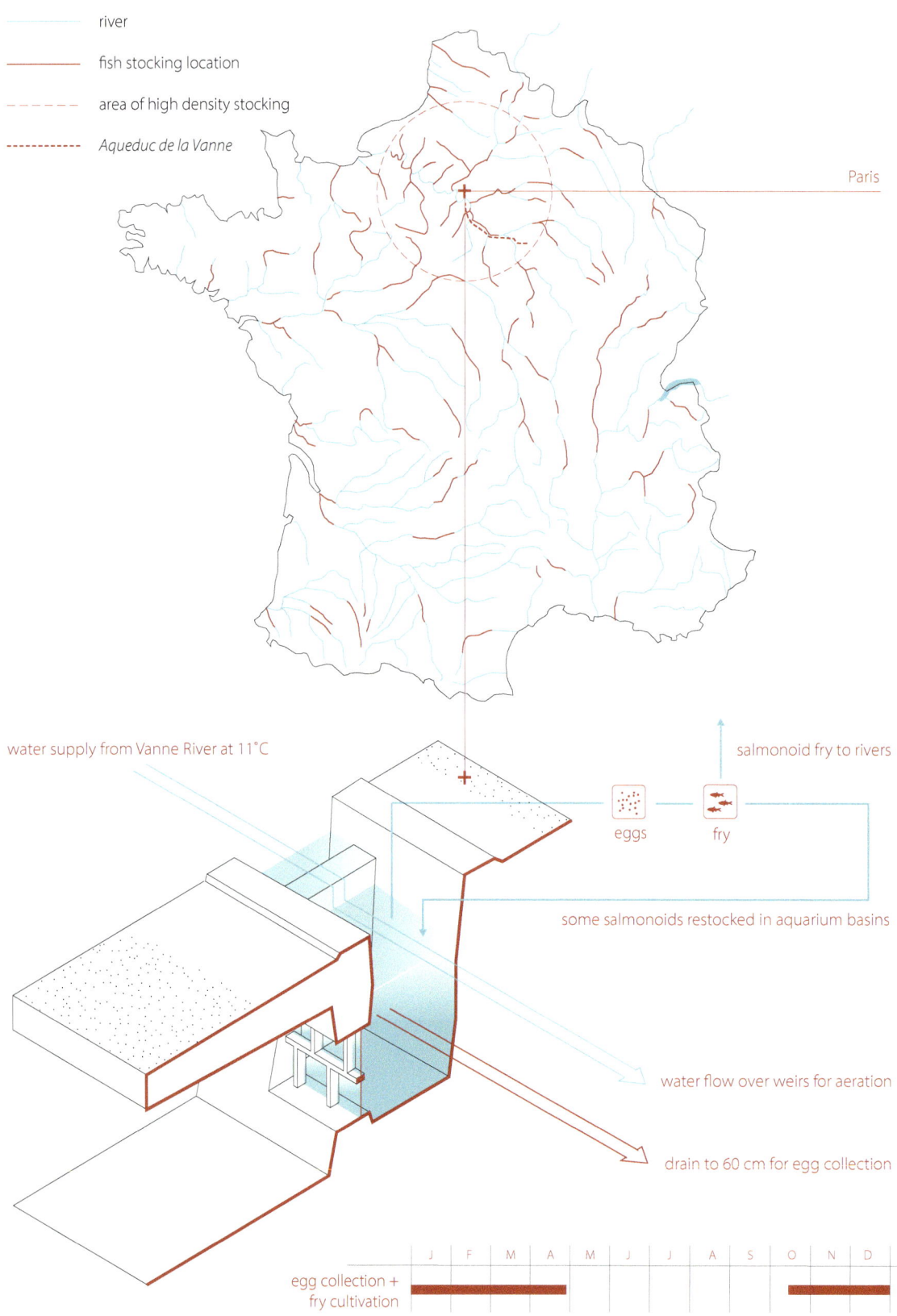

river

fish stocking location

area of high density stocking

Aqueduc de la Vanne

Paris

water supply from Vanne River at 11°C

salmonoid fry to rivers

eggs   fry

some salmonoids restocked in aquarium basins

water flow over weirs for aeration

drain to 60 cm for egg collection

	J	F	M	A	M	J	J	A	S	O	N	D
egg collection + fry cultivation												

08 *Landscape Systems and Strategies.* In the late nineteenth century the Aquarium du Trocadéro became a hub of aquaculture production and education in France. Basins were drained to facilitate egg collection, and salmonoid fry produced here were distributed to stock a number of rivers across France.

09 Agence de presse Meurisse, *Jardin du Trocadéro, vue de l'aquarium (intérieur)*, 1928. The open-air basins acted as both display tanks and light wells for the subterranean gallery.

01 The village of Ganvié at Lake Nokoué is populated by thousands of Tofino people, and thousands of tilapia. Aerial view of acadjas and houses reveals that implanting wood branches and piles into the lakebed is a strategy both for creating habitat for fish and for constructing housing raised above the lake.

Acadjas at Lake Nokoué, Benin

Landscape Type: acadja

Landscape Area: 400 ha (1981)

Aquaculture Yield: 8,000 kg/ha/yr (large acadja)
                   28,000 kg/ha/yr (small acadja)

Aquaculture Type: extensive

Water Type: brackish

Brush parks are a form of aquaculture constructed in coastal lagoons and rivers in areas of Africa, Asia, and South America. At brush parks, fish aggregation is induced by implanting tree branches in rivers, lakes and lagoons, and these vegetal constructions provide fish with food, shelter, and in some cases, space for reproduction. Given this mode of construction, brush parks tend to attract local fish species normally associated with vegetation and woody debris in rivers and lagoons. Although there are many forms, varieties, and sizes of brush parks, they typically consist of an inner core of densely packed softwood branches surrounded by an outer framework of hardwood branches. Some of the most sophisticated iterations of brush parks in West Africa are those constructed by the Tofinu people in Benin on Lake Nokoué.[1] Known locally as *acadjas* and ranging in size from 3 m² to 4 km², this type of aquaculture has been evolving for over a century.

The most complex versions of acadjas on Lake Nokoué are constructed with upright hardwood branches that form the peripheral structure and softwood branches both implanted upright and laid flat at the bottom of the acadja interiors. The underwater brushwood gradually decays, creating a substrate for periphyton, algae, and insect larvae, which are then consumed by a range of fish species including Blackchin Tilapia (*Sarotherodon melanotheron*). Acadjas that are fished shortly after installation of branches can be considered fish traps, while those left for longer periods between harvest serve as habitats where fish live and breed.[2]

To establish 1 ha of an acadja, softwood branches are planted at a density of 12 to 16 branches per square meter, equivalent to 40 t/ha dry weight of wood.[3] Due to decay, wood replacement rates in acadjas range from 30% to 75% per year. Yields from acadjas are higher when the density of interior branches is greater, and as a result this form of aquaculture has been a cause of deforestation in upland-wood production zones that are expanding northward. Unsustainable forest management in the Lake Nokoué watershed has increased siltation in the lake, which has been detrimental to acadja productivity.[4] The hundreds of tree branches planted in an acadja eventually accumulate silt and the acadja is colonized by vegetation; over time acadjas transition into islands for crop farming.[5] At Ganvié, aquaculture both feeds and forms an evolving village.

A high density of acadjas may be found in and around the village of Ganvié on the west coast of Lake Nokoué. This village was established in the sixteenth or seventeenth centuries and currently is home to approximately 20,000 Tofinu people and thousands of tilapia. Water is pervasive at the road-less village of Ganvié; it flows under colorful houses raised on wood piles, through the acadjas, and along village canals.

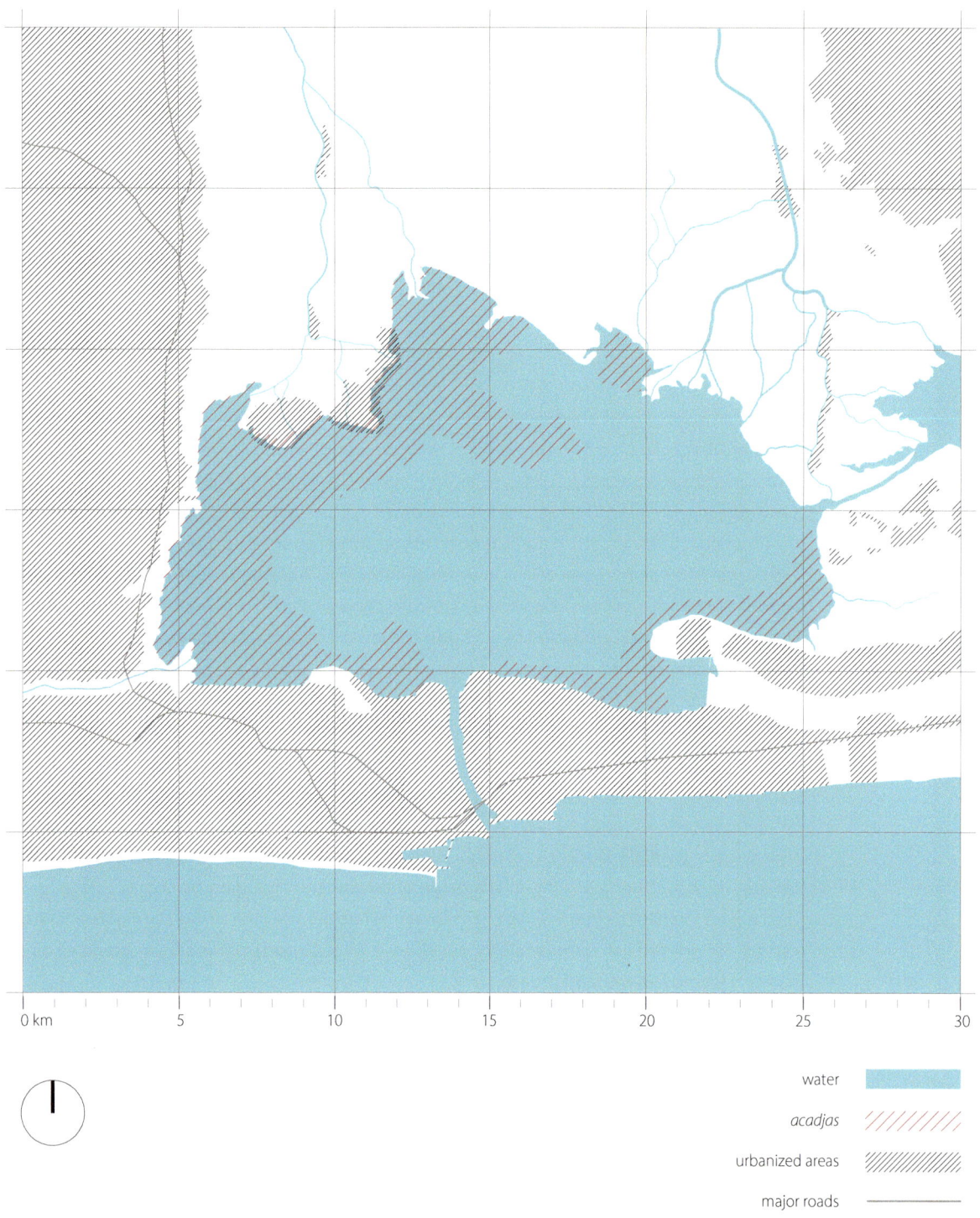

| water |
| acadjas |
| urbanized areas |
| major roads |

02 *Regional Diagram*. Lake Nokoué, a brackish lake that covers approximately 15,000 ha, is the largest and most productive brackish water body in Benin. The density of acadjas is highest on the west side of the lake, adjacent to the village of Ganvié. Cotonou, the largest city in Benin, lies between Lake Nokoué and the Atlantic Ocean.

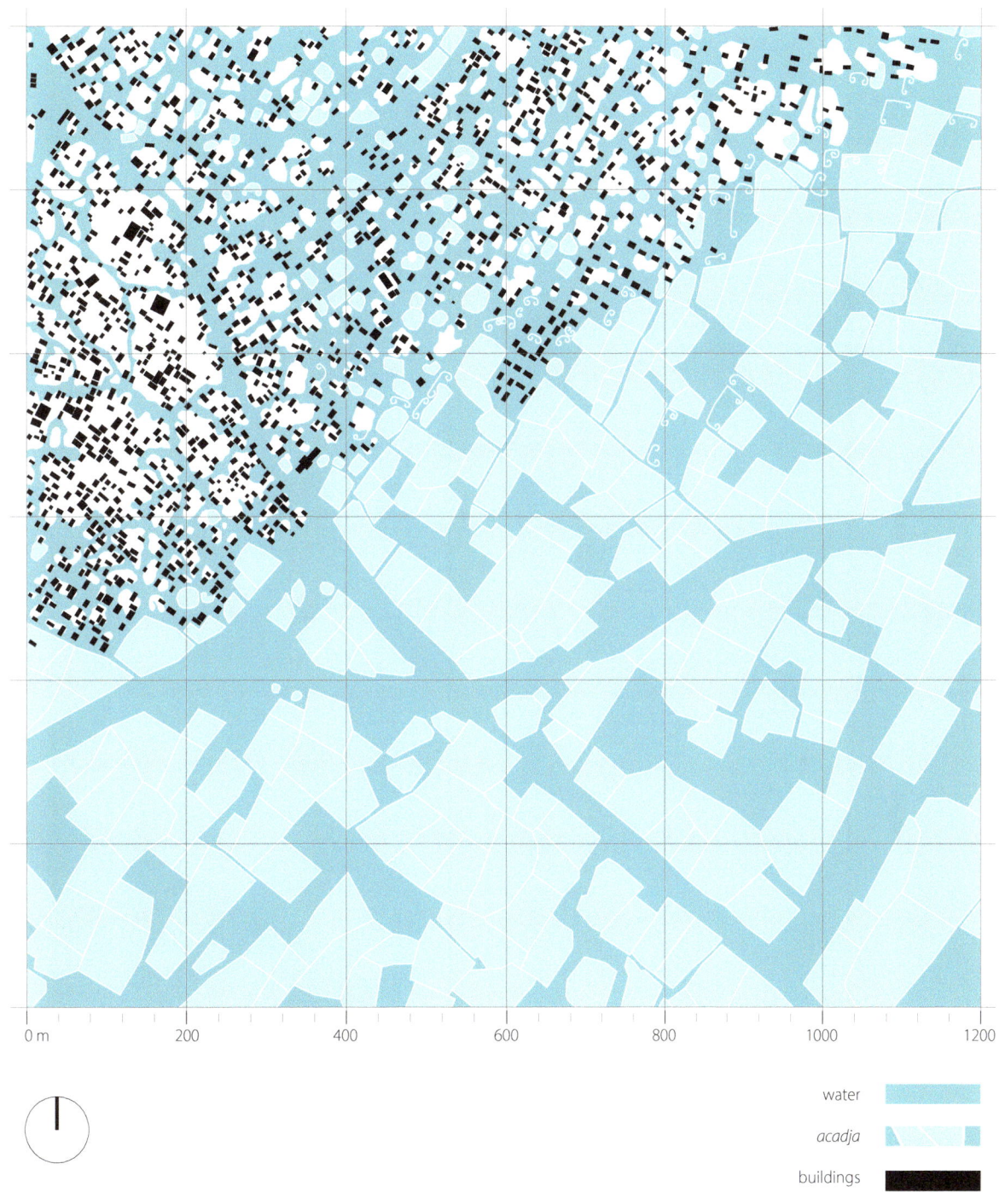

water
acadja
buildings

0 m · 200 · 400 · 600 · 800 · 1000 · 1200

03 *Site Diagram.* At the lakeside village of Ganvié, the boundary between settlements and aquaculture is blurred. Acadjas infiltrate the aggregation of buildings, and canals provide boat access to both the road-less village and the field of fish pens.

Cultivated Fauna

Blackchin Tilapia *(Sarotherodon melanotheron)*
catfish *(Chrysichthys spp.)*

Cultivated Flora

oil palm *(Elaies guineense)*
ironwood *(Casuarina equisetifolia)*
mangroves *(Avicennia africana
and Laguncularia racemosa)*
ackee *(Blighia sapida)*
water lettuce *(Pistia stratiotes)*

island

house constructed on wood piles

hardwood branches at *acadja* perimeter

softwood branches at *acadja* interior

40

0m

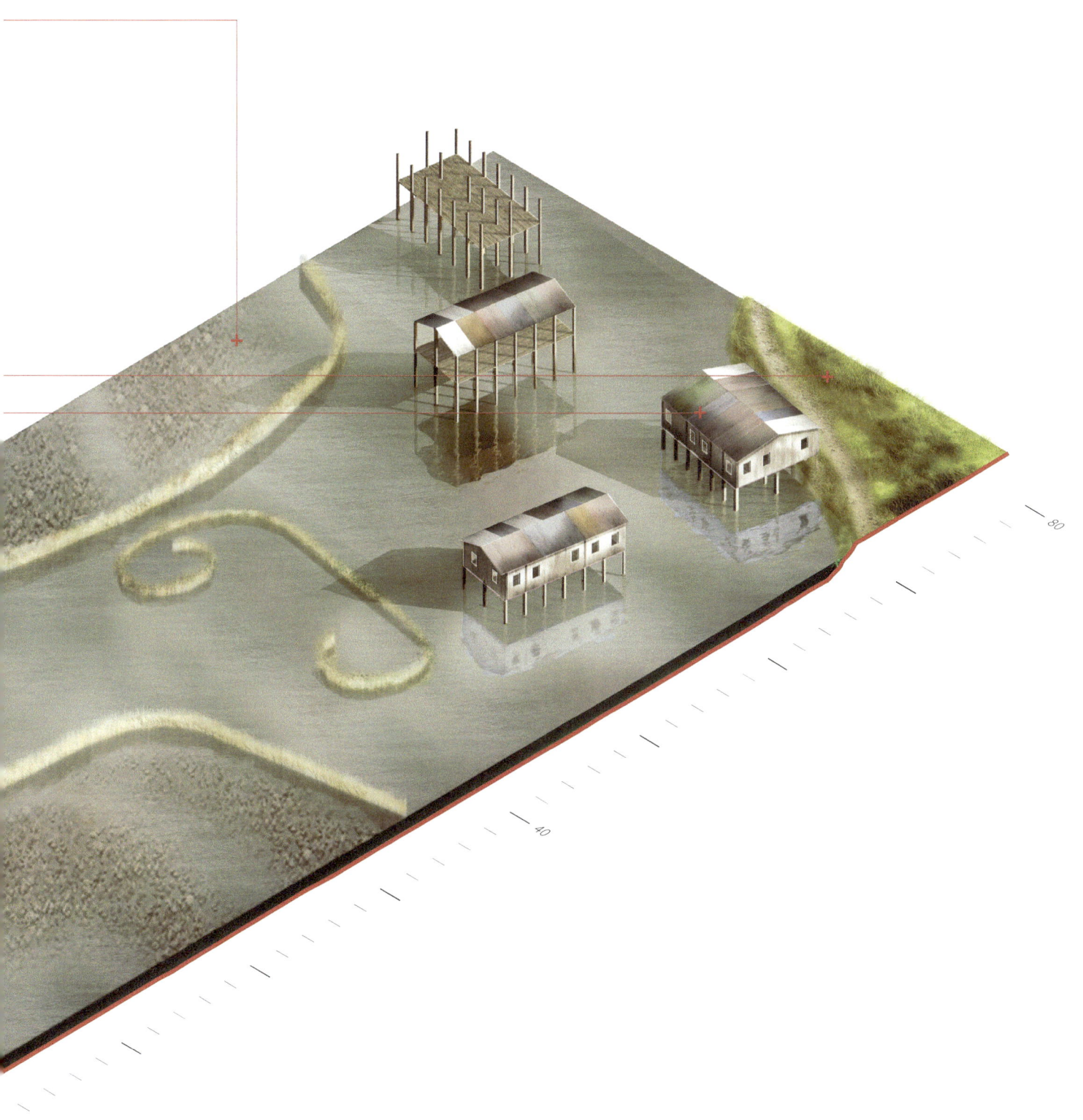

04 *Transect*. Wood implanted in the lake bottom structures both housing and aquaculture at Lake Nokoué. Houses are raised on wood piles and clad in salvaged sheet metal and plywood. Implanted hardwood branches form the perimeters of the acadjas, while softwood branches at acadja interiors are approximately 2.5 m tall and buried 50 cm in the lake bottom.

05 Aerial view of acadjas, canals, and houses at Ganvié. The decaying softwood at the interior of each acadja attract fish such as tilapia and catfish seeking refuge zones and feeding grounds.

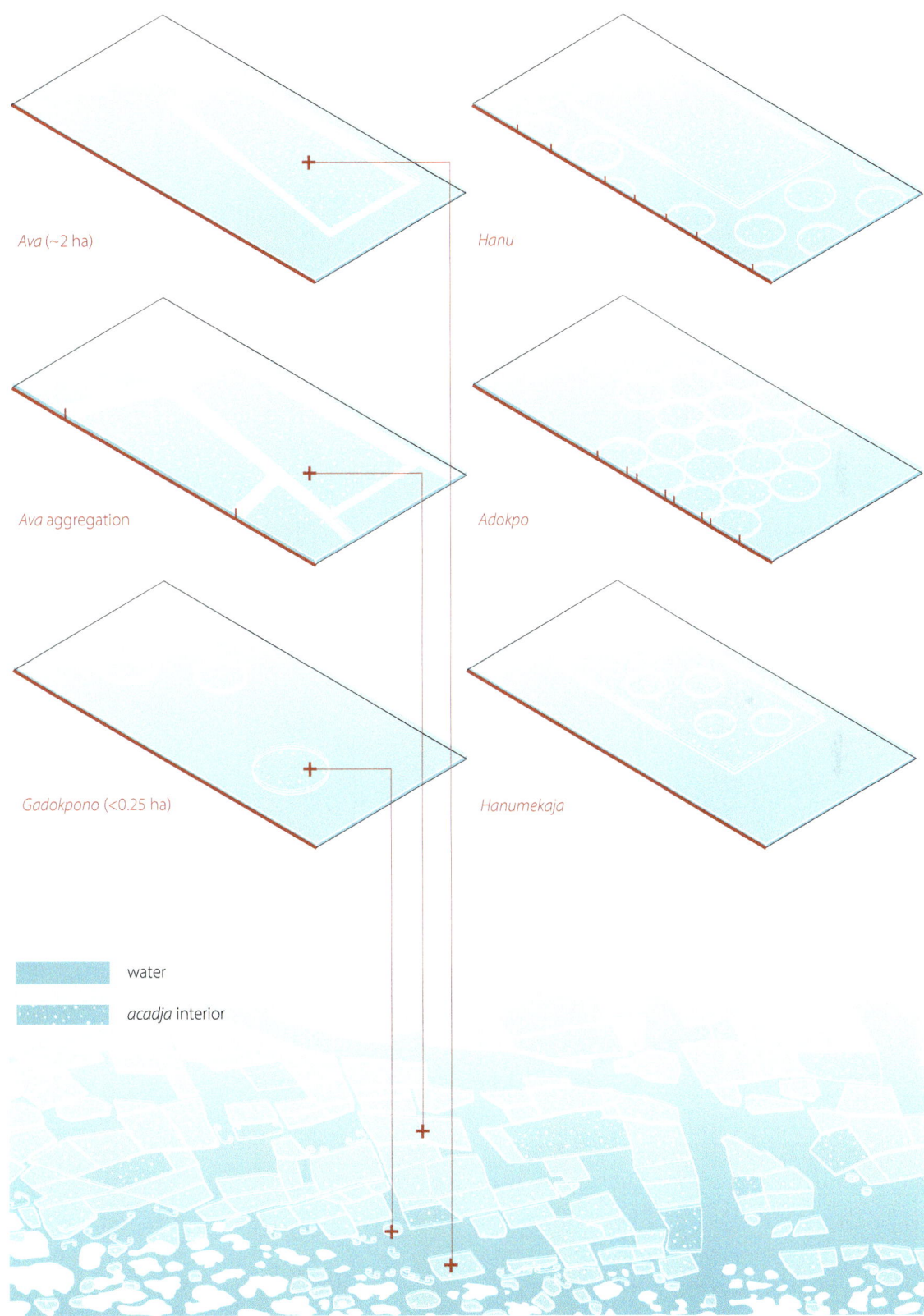

Ava (~2 ha)

Hanu

Ava aggregation

Adokpo

Gadokpono (<0.25 ha)

Hanumekaja

water

acadja interior

06 *Landscape Systems and Strategies.* There are a range of acadja types across Lake Nokoué, and the types differ in shape, size, harvest procedures, and harvest rates. At Lake Nokoué, the acadja system is predominantly characterized by a type known locally as Ava.

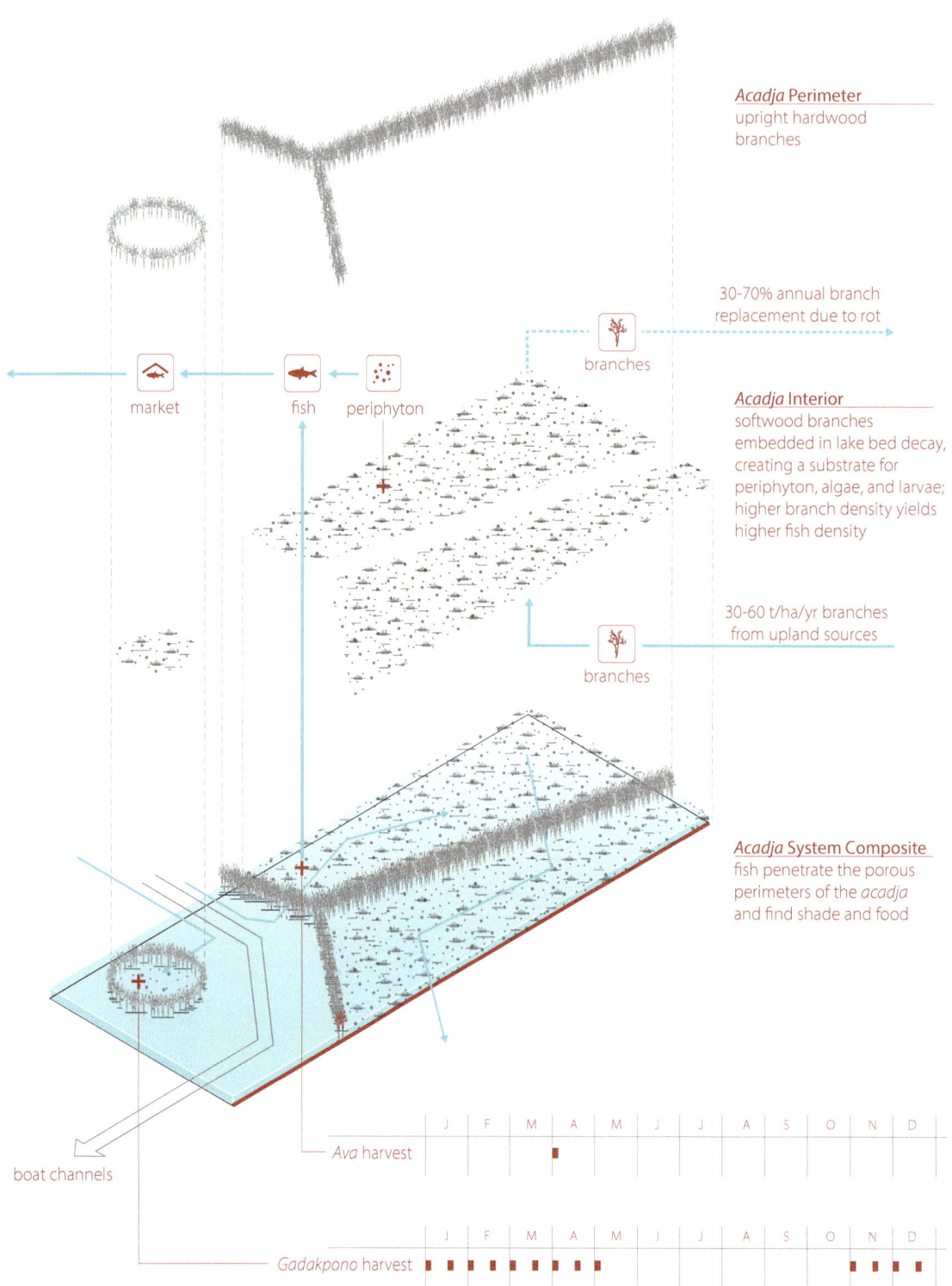

Acadja Perimeter
upright hardwood
branches

30-70% annual branch
replacement due to rot

branches

Acadja Interior
softwood branches
embedded in lake bed decay,
creating a substrate for
periphyton, algae, and larvae;
higher branch density yields
higher fish density

30-60 t/ha/yr branches
from upland sources

branches

Acadja System Composite
fish penetrate the porous
perimeters of the *acadja*
and find shade and food

market   fish   periphyton

boat channels

	J	F	M	A	M	J	J	A	S	O	N	D
Ava harvest				■								

	J	F	M	A	M	J	J	A	S	O	N	D
Gadakpono harvest	■ ■	■	■	■	■					■	■	■

07 *Landscape Systems and Strategies*. Acadjas are characterized by the mix of hardwood and softwood branches utilized in the construction and operation of these systems. The yield of an acadja increases logarithmically as the density of planted softwood branches increases. The replacement rate of softwood at acadjas is high due to decay.

08 View to houses standing within a field of acadjas.
09 Transport of hardwood bundles for construction of acadjas.

01 At the East Kolkata Wetlands (EKW), the bheris, earthen basins where fish are raised and harvested, are part of a landscape-based system of treating and upcycling municipal wastewater. This system supports aquaculture, agriculture, and provides habitat for diverse flora and fauna.

Bheries of the East Kolkata Wetlands, India

Landscape Type: bheri

Landscape Area: 5,842 ha (bheris, 2019) | 12,500 ha (EKW, 2019)

Aquaculture Yield: 6,480 kg/ha/yr (2019)

Aquaculture Type: extensive/ semi-intensive

Water Type: fresh

The East Kolkata Wetlands (EKW) is a 12,500 ha productive wetland constructed on the eastern edge of the city of Kolkata, in India. Dhrubajyoti Ghosh, the celebrated ecologist recognized for conservation of the EKW, describes a "lasting tradition of disposal and utilization of urban waste in agriculture and fisheries. The locals here have developed a remarkable system to help meet the three basic problems of developing countries: shortage of food, shortage of employment opportunities, and shortage of funds to treat waste."[1] In addition to being one of the largest sewage-fed aquaculture systems in the world, the wetland features paddy agriculture, solid waste disposal, habitation, markets, and transportation systems. Approximately 90% of the 8,500 people who are employed in this aquaculture project reside in the EKW.[2]

Near the end of the nineteenth century, the area of the future EKW was transformed from brackish lakes to freshwater-wetland disposal infrastructure by means of constructed canals, sluices, and dikes that directed municipal wastewater and stormwater through the wetlands. Experimentation with sewage-fed aquaculture at the EKW began in 1929. The area utilized for fish farming has declined over time, from a high of 7,300 ha in 1945 to 5,800 ha in 2003.[3] The reduction is due in large part to the conversion of what the locals called *bheris* (fishponds) to agriculture paddies. In 2010, the 264 bheris of the EKW produced over 15,000 mt of fish per year and fresh fish were sold daily at the seven auction markets within and adjacent to the EKW.[4]

Municipal wastewater management in the EKW is a complex choreography of material movement and nutrient cycling across a gradient of wet-to-dry landscapes, including fishponds, shallow rice paddies, and terrestrial fields. The process begins with the mechanical pumping of wastewater through canals to anaerobic-wetland basins where pathogens are removed. Solids in these ponds are then dredged and composted to fertilize agricultural fields that are colocated with a municipal solid-waste landfill. The water then flows to maturation ponds for further processing and pathogen removal, a process that sponsors the growth of algae and duckweed that is in turn supplied as feed for fish. The water finally flows into bheris, where several species of fish are stocked, raised, and harvested. The water from fish farming, high in nitrogen and phosphorous, is a resource for agriculture and is routed to fertilize adjacent rice paddies.[5]

The EKW provides the region with a range of ecosystem services and was named a RAMSAR Wetland of International Importance in 2003. However, land-use changes, increased siltation, and industrial pollution have led to a decline in biodiversity that threatens the ecological function of the system.

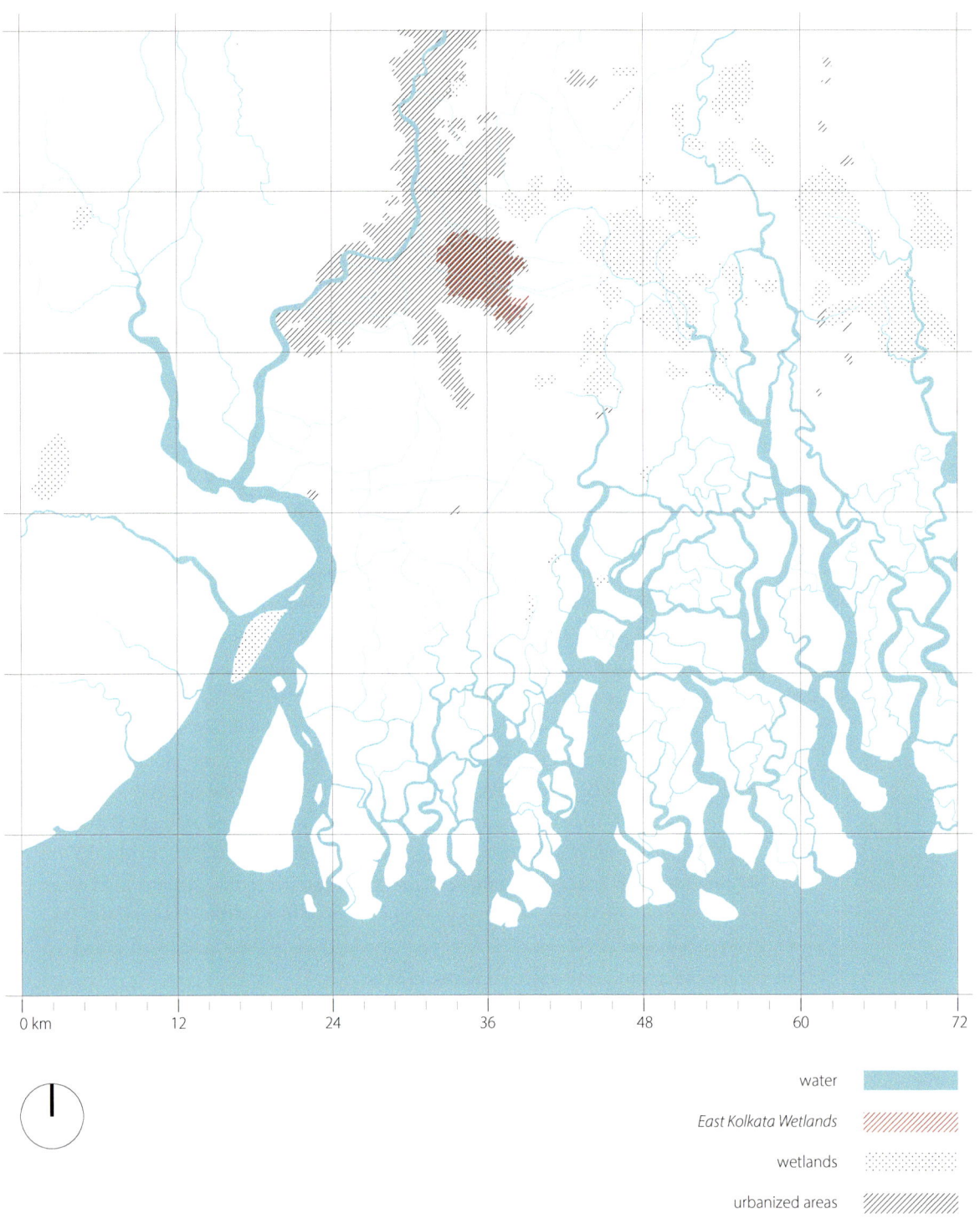

water	
East Kolkata Wetlands	
wetlands	
urbanized areas	

02 *Regional Diagram.* The EKW, located on the eastern fringes of Kolkata, occupy 12,500 ha. The EKW is one of the largest assemblages of sewage-fed fish ponds in the world. These wetlands are part of the extensive inter-distributary wetland regimes formed by the Gangetic Delta.

water	
buildings	
roads	
trails	
water hyacinth	
wastewater inlet	<

0 m   200   400   600   800   1000   1200

03 *Site Diagram.* Fish in the EKW are reared in bheries that are supplied with municipal wastewater. Approximately 8,500 people are directly involved with sewage-fed aquaculture in the EKW, and 90% of this population reside in settlements constructed on wide dikes that wind across the wetlands.

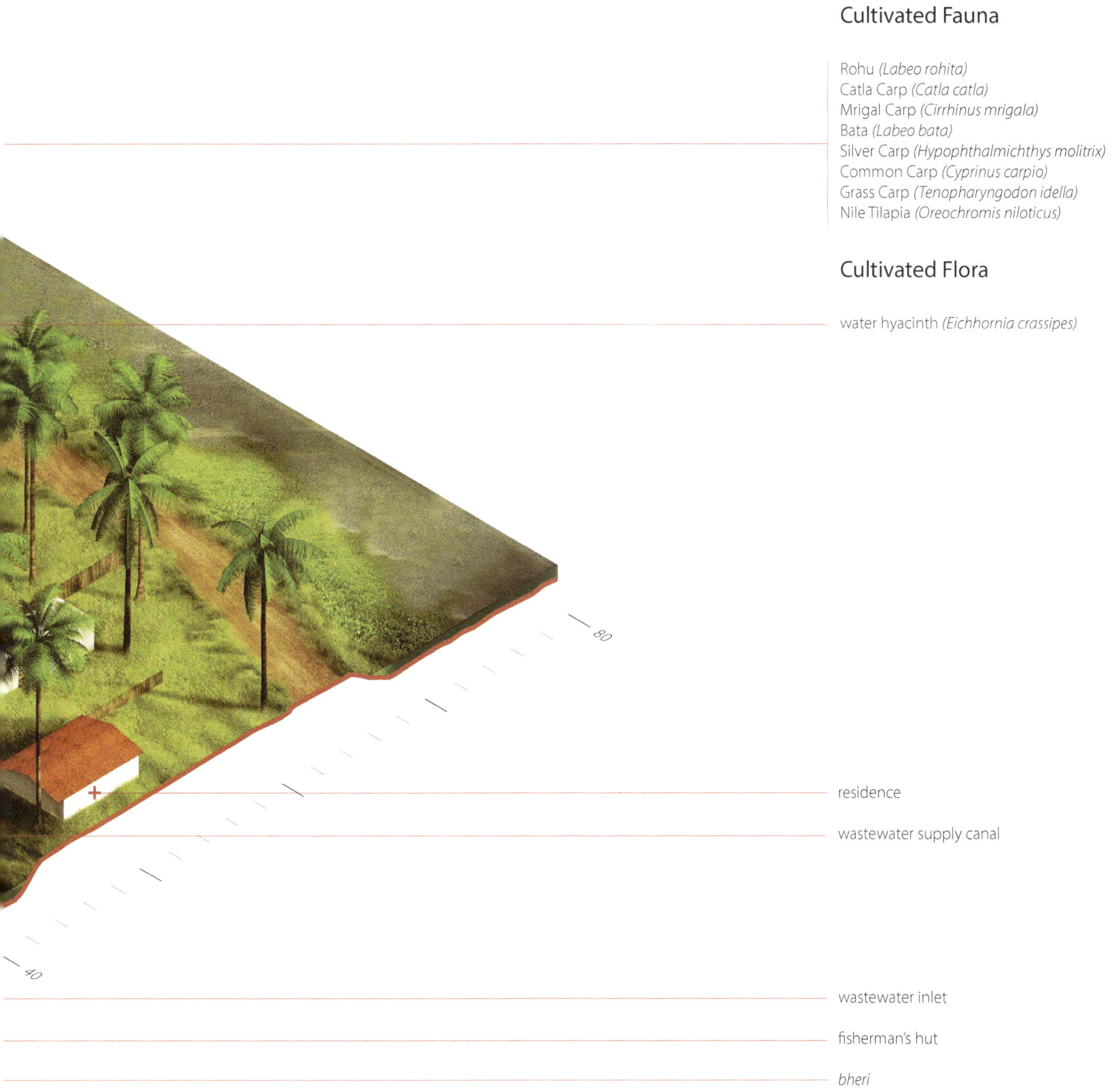

Cultivated Fauna

Rohu *(Labeo rohita)*
Catla Carp *(Catla catla)*
Mrigal Carp *(Cirrhinus mrigala)*
Bata *(Labeo bata)*
Silver Carp *(Hypophthalmichthys molitrix)*
Common Carp *(Cyprinus carpio)*
Grass Carp *(Tenopharyngodon idella)*
Nile Tilapia *(Oreochromis niloticus)*

Cultivated Flora

water hyacinth *(Eichhornia crassipes)*

residence

wastewater supply canal

wastewater inlet

fisherman's hut

bheri

04 *Transect.* Canals distribute municipal waste water to bheries across the EKW, and each bheri dike has a culvert inlet and sluice to control waste water flow into the basin. Water depth in the ponds varies from 0.5 m to 1.5 m.

05 Fish harvest at a bheri. Fish are loaded into baskets and within two hours are carried to local fish markets by foot or bicycle; there is no system of refrigeration or use of ice in the harvest and sale process.
06 The sewage water feeder canal in the foreground supplies bheries of various size and function in the distance.

Aquaculture +
Hydrologic Flows

- - - - - - - EKW boundary

· · · · · · · drainage basin

bheris

——————— drainage canal

——————— *bheri* canal

▲ pump station

🐟 live fish markets

Hooghly River

to Kulti River

Site Composite

- - - - - - - EKW boundary

bheris

paddy agriculture

landfill agriculture

——————— canal

——————— road

07 *Landscape Systems and Strategies.* Municipal sewage is upcycled and used in a sequence of economic activities: wastewater aquaculture in over 260 bheries, vegetable farming on a municipal landfill, and paddy cultivation using fishpond effluent. Other aquaculture infrastructural elements within and around the EKW include fish markets and nurseries.

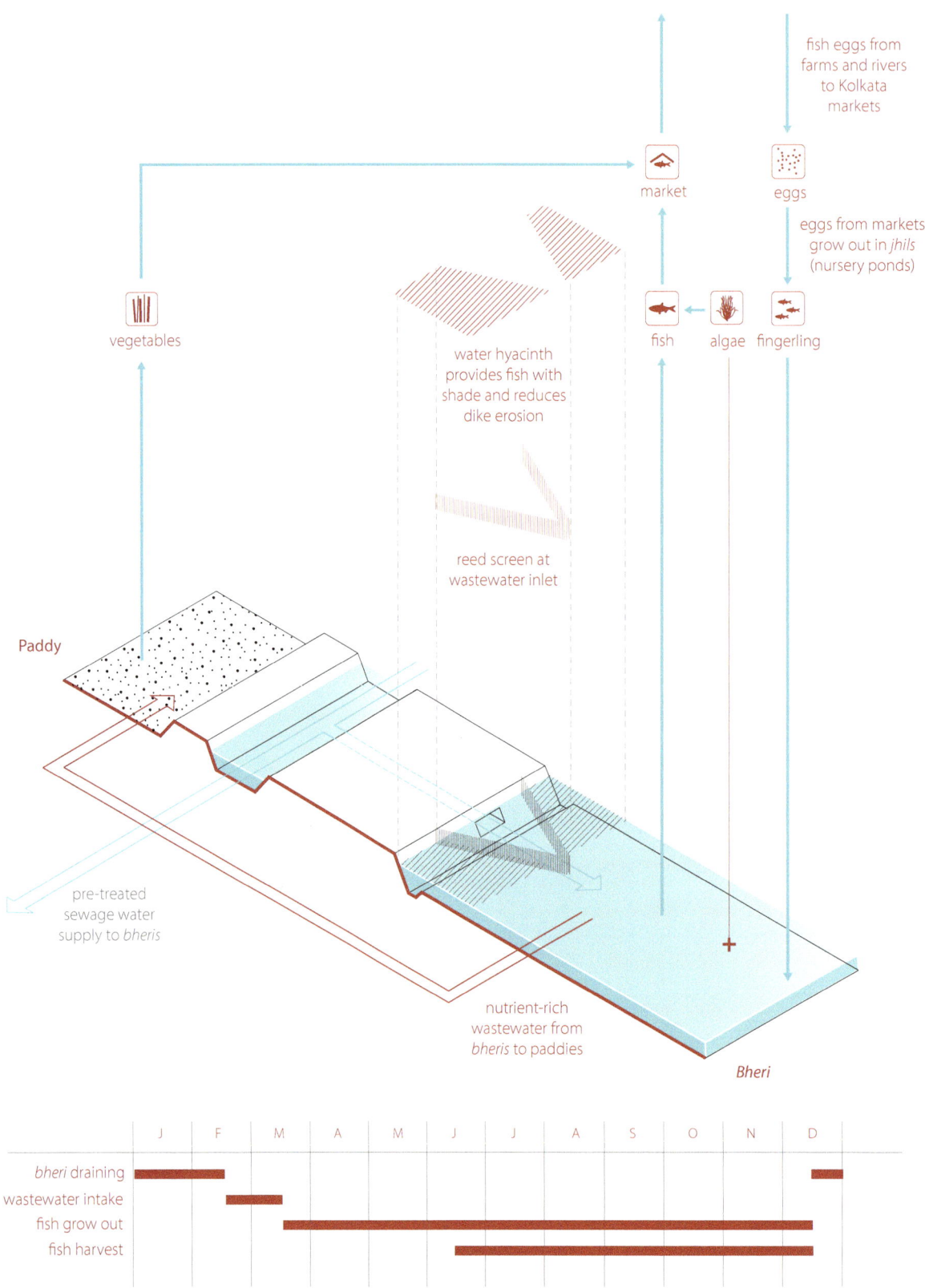

fish eggs from
farms and rivers
to Kolkata
markets

market   eggs

eggs from markets
grow out in *jhils*
(nursery ponds)

vegetables

water hyacinth
provides fish with
shade and reduces
dike erosion

fish   algae   fingerling

reed screen at
wastewater inlet

Paddy

pre-treated
sewage water
supply to *bheris*

nutrient-rich
wastewater from
bheris to paddies

Bheri

	J	F	M	A	M	J	J	A	S	O	N	D
bheri draining	▬▬											▬
wastewater intake		▬▬										
fish grow out			▬▬▬▬▬▬▬▬▬▬▬▬▬▬									
fish harvest						▬▬▬▬▬▬▬▬▬▬▬						

08 *Landscape Systems and Strategies.* A generalized calendar of aquaculture activities at larger bheries in the EKW includes annual drainage, wastewater intake, and fish grow out and harvest. Bheries are typically lined with a 4 m band of water hyacinth, which serves as a source of food and shade for fish and also protects earthen dikes from erosive waves.

09 Fishing shack standing within the band of water hyacinth lining a bheri.
10 Wastewater inlet at a bheri. The triangular reed screen prevents clogging due to vegetation growth and sedimentation.

01 Contemporary satellite image of Shunde District in the city of Foshan, Guangdong Province, illustrates the juxtaposition of old and new patterns of aquaculture and settlement typical of the rapidly urbanizing Pearl River Delta (PRD).

Dike-Pond System of the Pearl River Delta, China

Landscape Type: dike-pond system

Landscape Area: 56,301 ha (1988)

Aquaculture Yield: 7,000 kg/ha/yr (1988)

Aquaculture Type: extensive

Water Type: fresh

The dike-pond system (DPS) of the Pearl River Delta (PRD) in South China integrated aquaculture and agriculture. The system paired carp polyculture in ponds with a variety of terrestrial crops grown on adjacent dikes to create a complex, constructed ecosystem. At its peak development in the late-twentieth century, saleable outputs from the DPS included silk, vegetables, sugar cane, fruit, and flowers, along with up to five varieties of carp that constituted 80% of China's live-fish exports.[1] Writing in 2002, scholars Kenneth Ruddle and Gongfu Zhong describe the DPS as polyculture which was "operated on a geographical and economic scale unrivaled elsewhere in the world."[2]

During the Ming Dynasty (1368-1644) the DPS emerged as a strategy for occupying flood-prone delta landscapes by means of pond excavation and the construction of arable dikes. The mulberry DPS became the dominant iteration of the system in the seventeenth century. In this system, mud from the bottom of the ponds was excavated two to three times annually and spread on top of adjacent dikes to serve as nutrient-rich fertilizer that supported the growth of mulberry trees. Leaves from the mulberry trees were harvested as feed for silkworms. The worm excrement returned to the pond as food for the carp while wastewater from silk factories was also cycled back to the ponds. Due to the system's profitability, the mulberry DPS expanded to its peak in 1925, when 93,000 ha of dikes in the PRD were planted with mulberry.[3] In addition to managing flood water in the delta, the DPS was also ecologically beneficial to the larger region by cycling water and mud in the fishponds that was high in nitrogen and phosphorous, rather than allowing nutrients to pollute the estuary.

The mulberry DPS reached its most advanced stage of ecological complexity in the 1980s. Since then, however, the system has declined due to the pressures of rapid urbanization and changes in the aquaculture-agriculture market. The introduction of the market-driven economy in China, beginning in the late 1970s, replaced collective ownership of agriculture. This led to the de-linking of agriculture and aquaculture. Today, monoculture of species like eel and prawn is more lucrative than dike cropping, and the dikes are being remade narrower to maximize pond surface area.[4]

Rapid urban expansion and new aquaculture regimes must negotiate the existing, prevalent aquaculture matrix. Some new development projects retain the centuries-old ponds and dikes, while other projects simply fill and erase the existing topography. A considerable amount of soil is needed to reclaim these deep ponds for urban development, and this has resulted in the excavation of local hills and mountains that centuries ago were islands in the PRD.

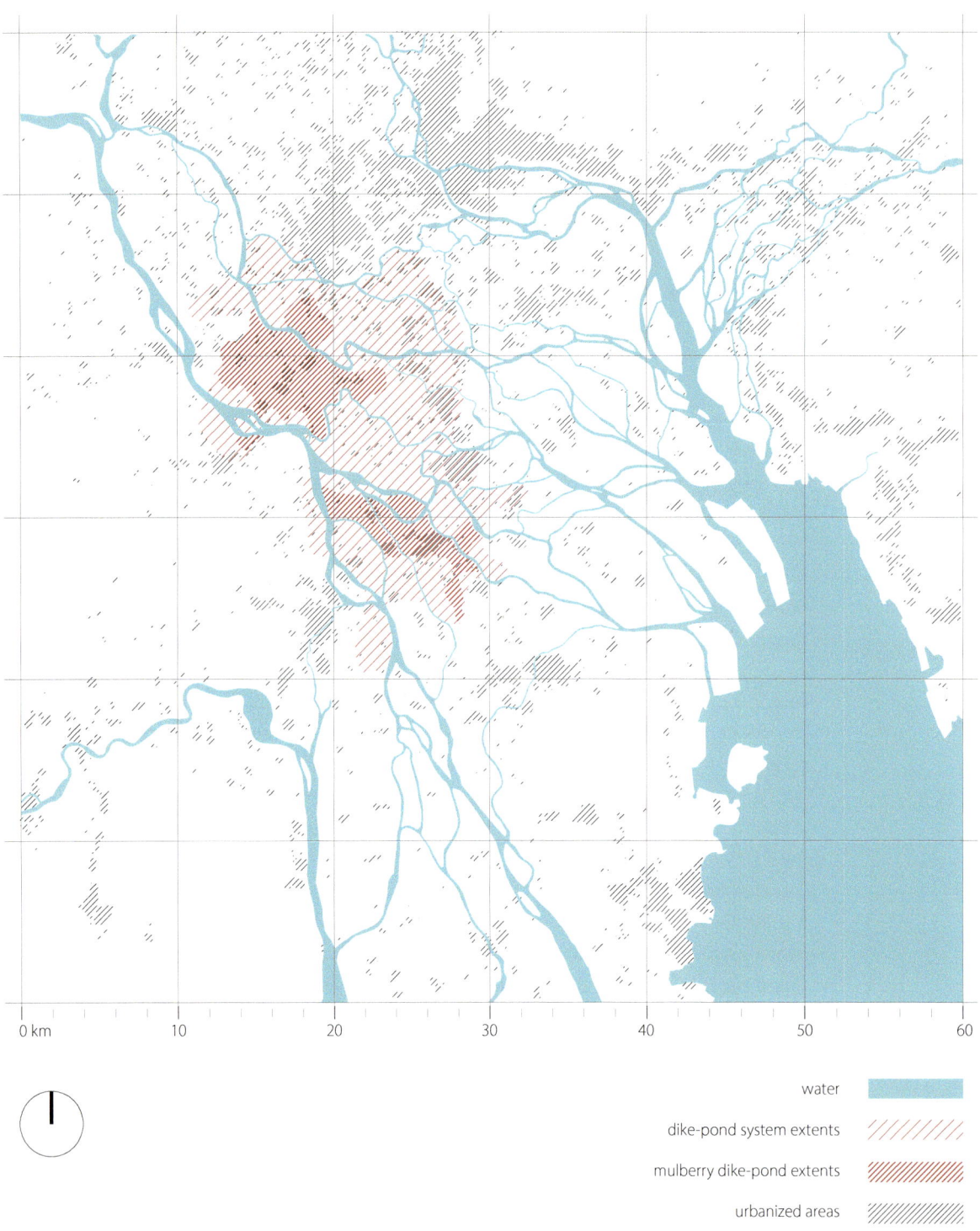

0 km	10	20	30	40	50	60

water

dike-pond system extents

mulberry dike-pond extents

urbanized areas

02 *Regional Diagram.* The contemporary landscape of the PRD began to take shape in the ninth century as earth from the excavation of fishponds in the delta was used to create a series of arable dikes. This diagram depicts the PRD in the mid-1980s, when the DPS was extensive and covered over 56,000 ha.

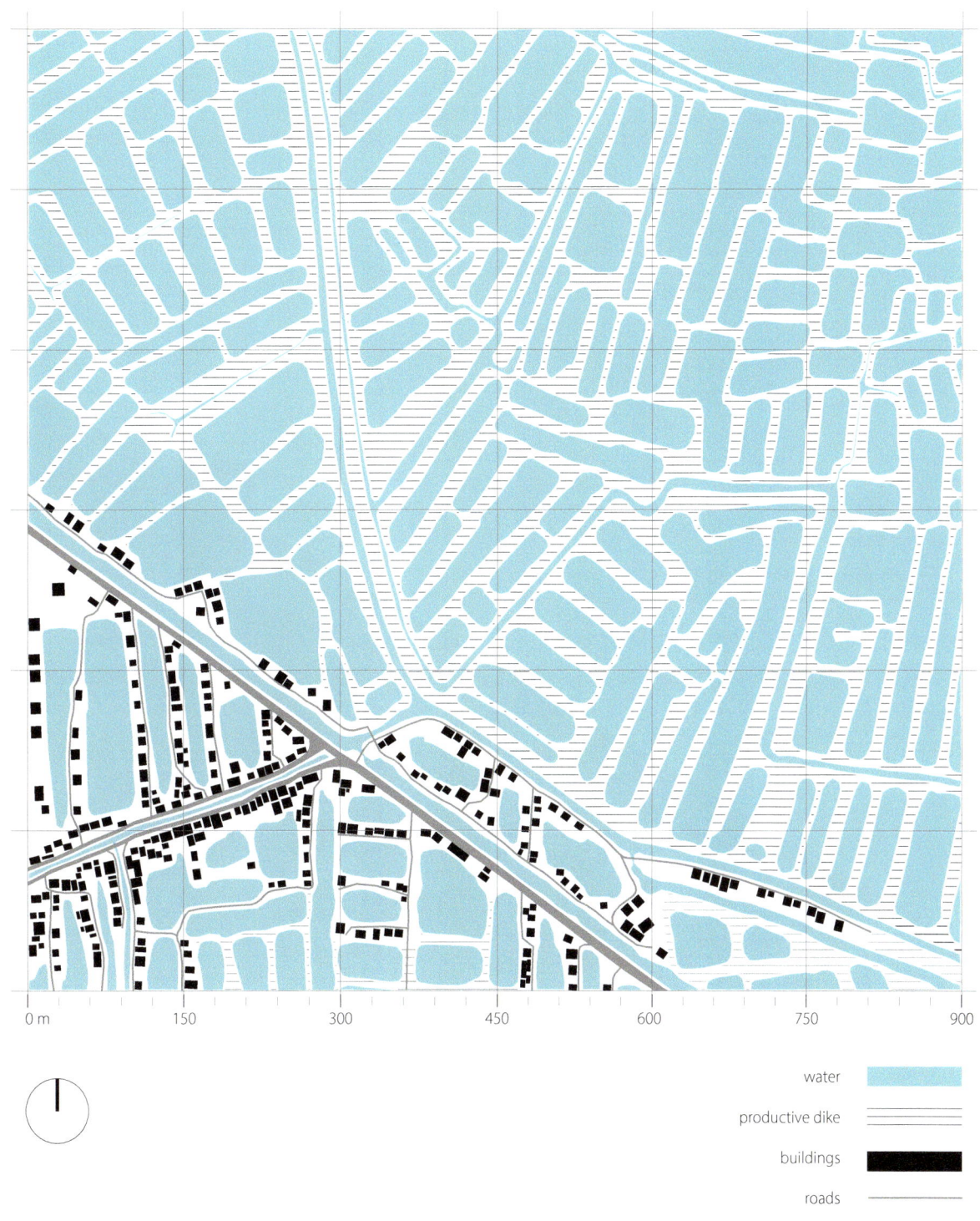

water

productive dike

buildings

roads

03 *Site Diagram*. Villages within the DPS, such as this one in Shunde District, Guangdong Province, traditionally took the form of housing constructed along dikes. In the twenty-first century, the fabric of the DPS is increasingly fragmented and most new urban development fills over pre-existing ponds and canals.

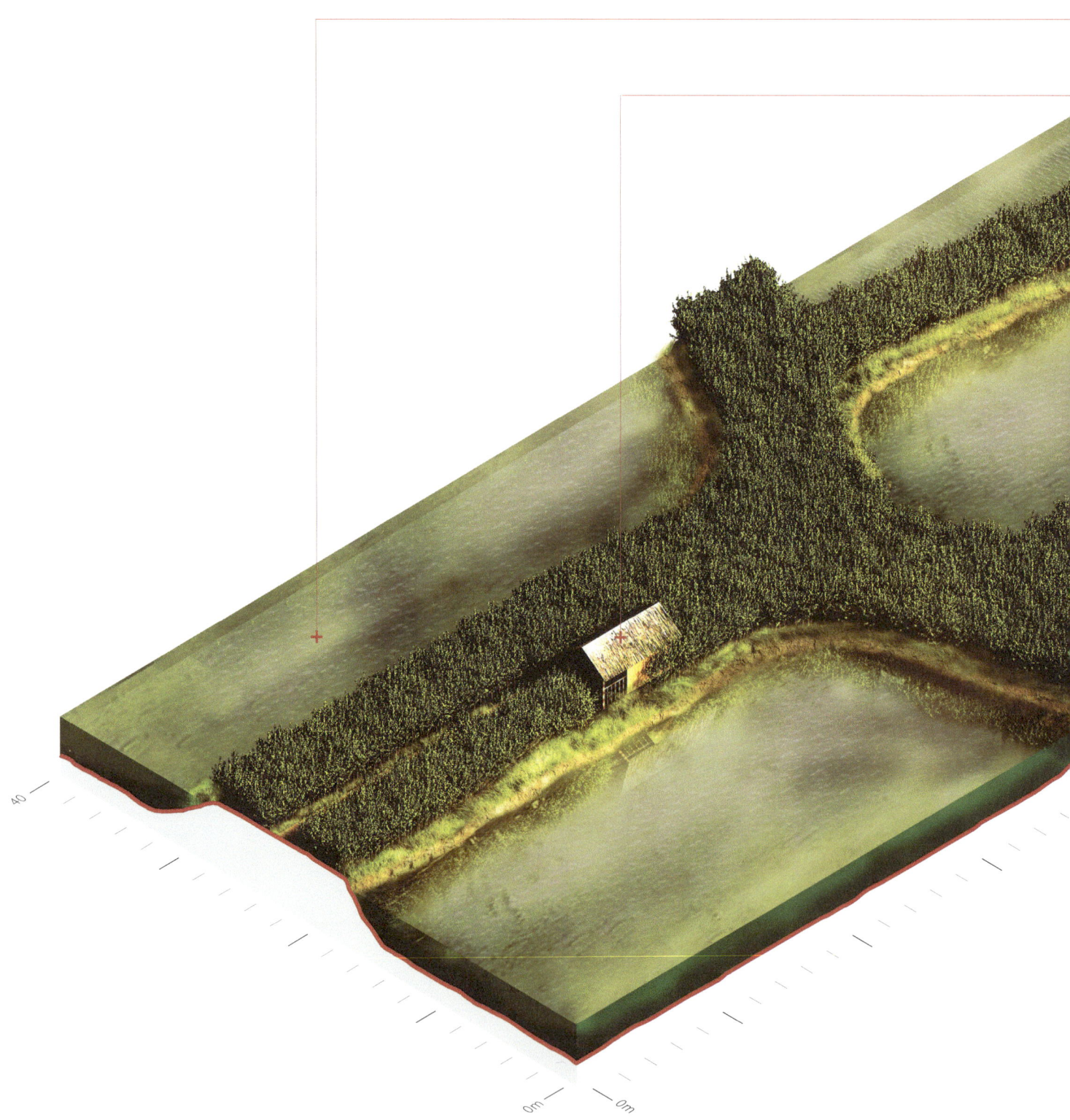

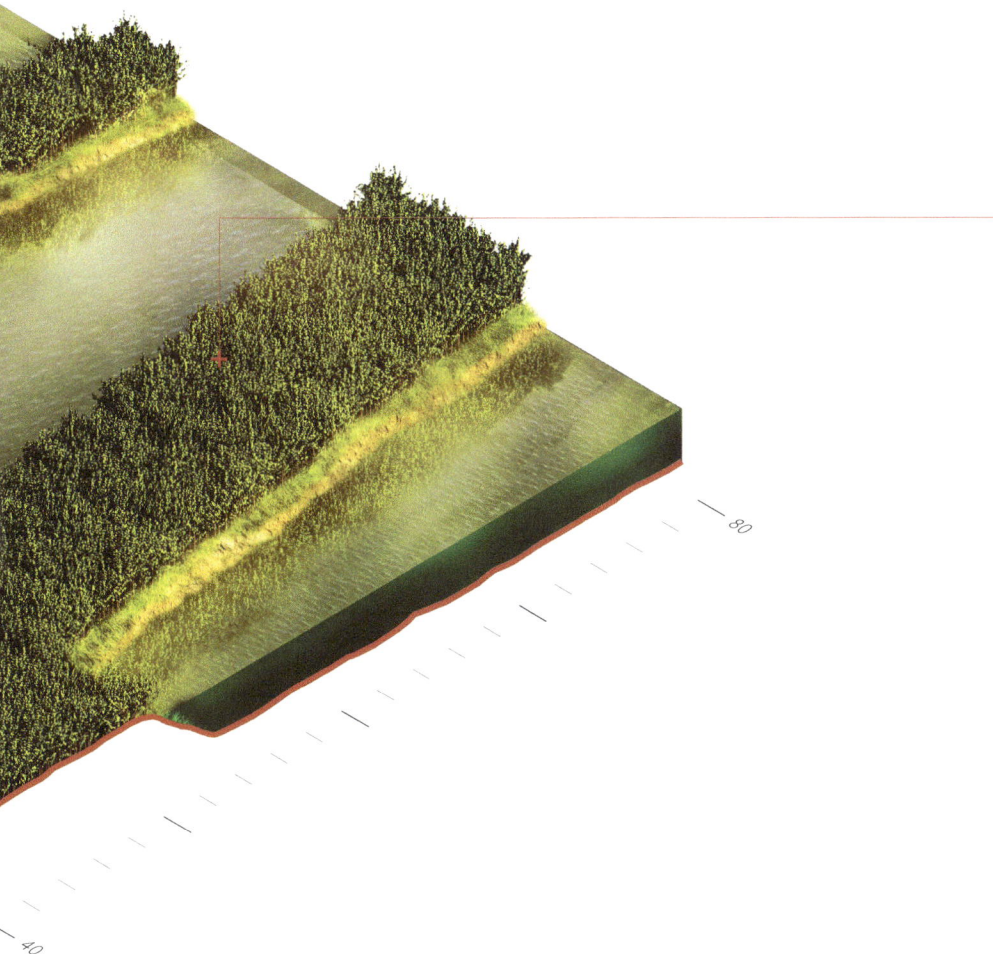

Cultivated Fauna

Bighead Carp *(Aristichthys nobilis)*
Silver Carp *(Hypophthalmichthys molitrix)*
Grass Carp *(Ctenopharyngodon idella)*
Common Carp *(Cyprinus carpio)*
Mud Carp *(Cirrhinus molitorella)*

silkworm *(Bombyx mori)*

pig *(Sus spp.)*
chicken *(Gallus gallus domesticus)*
duck (various)

Cultivated Flora

mulberry *(Morus atropurpurea)*

sugarcane *(Saccharum officinarum)*
banana *(Musa acuminata)*
flowers (various)
vegetables (various)

04 *Transect.* The mulberry dike-ponds depicted here illustrate a 60:40 pond to dike ratio. This proportion is based on optimizing the volume of pond mud that could be utilized as fertilizer at dikes that were between 6 m and 20 m wide. The ponds covered approximately 0.5 ha and had an average depth of 3 m, a depth that supported a polyculture of carp.

05 Aerial view of the DPS in the 1980s from Xiqiao Mountain in Nanhai District, Guangdong Province.
06 Harvest of mulberry leaves used to feed silkworms in the mulberry DPS.

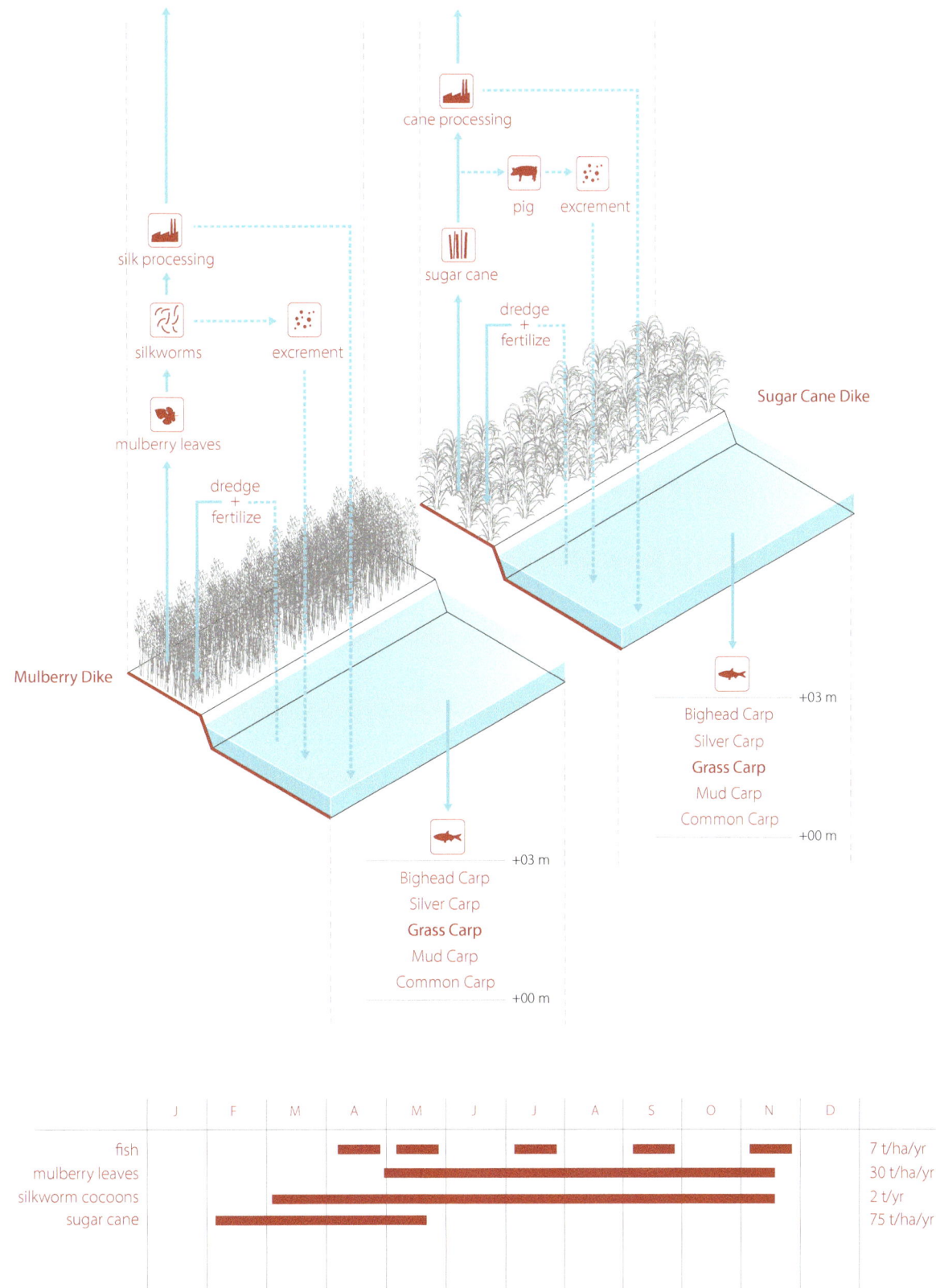

cane processing

silk processing

pig   excrement

silkworms   excrement

sugar cane

mulberry leaves

dredge
+
fertilize

dredge
+
fertilize

Sugar Cane Dike

Mulberry Dike

+03 m

Bighead Carp
Silver Carp
Grass Carp
Mud Carp
Common Carp
+00 m

+03 m

Bighead Carp
Silver Carp
Grass Carp
Mud Carp
Common Carp
+00 m

	J	F	M	A	M	J	J	A	S	O	N	D	
fish													7 t/ha/yr
mulberry leaves													30 t/ha/yr
silkworm cocoons													2 t/yr
sugar cane													75 t/ha/yr

07 *Landscape Systems and Strategies*. In the DPS, aquaculture, agriculture, and animal husbandry were integrated in a multispecies ecosystem. Nutrient and energy flow between the dike and the pond was facilitated through pond mud (which fertilized dike crops) and dike crops (which fed fish). At least four types of systems were developed in the PRD.

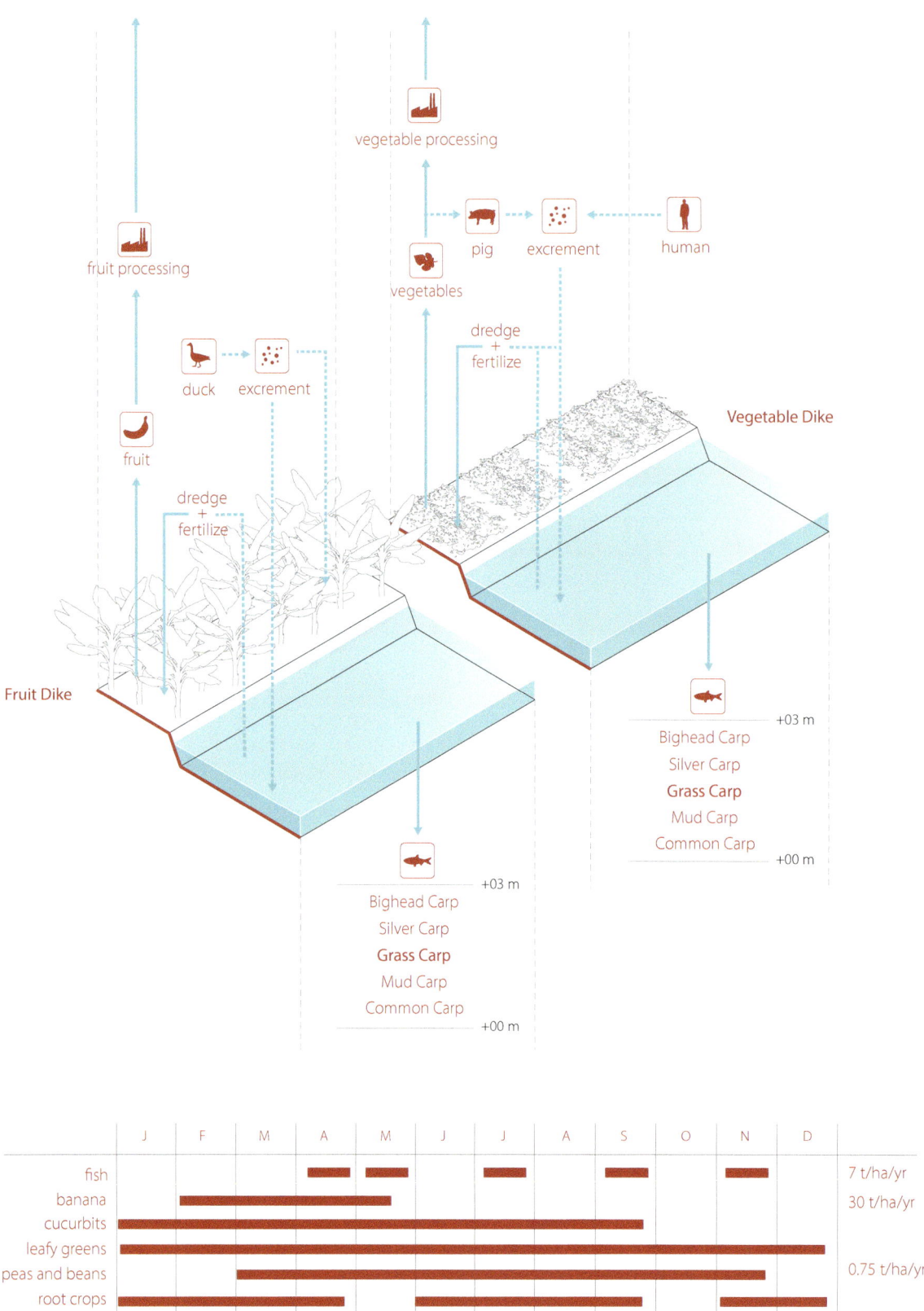

vegetable processing

fruit processing

pig **excrement** **human**

vegetables

duck **excrement**

dredge
+
fertilize

fruit

dredge
+
fertilize

Vegetable Dike

Fruit Dike

+03 m
Bighead Carp
Silver Carp
Grass Carp
Mud Carp
Common Carp
+00 m

+03 m
Bighead Carp
Silver Carp
Grass Carp
Mud Carp
Common Carp
+00 m

	J	F	M	A	M	J	J	A	S	O	N	D	
fish													7 t/ha/yr
banana													30 t/ha/yr
cucurbits													
leafy greens													
peas and beans													0.75 t/ha/yr
root crops													

08 *Landscape Systems and Strategies.* In the DPS at least five species of carp coexist in vertically striated aquatic niches in deep fishponds. Grass Carp, fed elephant grass and excrement produced at dikes, accounted for 50% of all fish production. Grass Carp waste fueled growth of plankton consumed by other carp species in the pond.

09 View from a vegetable dike to a fishpond.
10 Satellite photos over ten years of a site in Shunde District, Guangdong Province, depict the transformation of the irregular mosaic of DPS aquaculture into the regularized grid of contemporary, intensive monoculture fish farms.

01 Rice-fish terraces at Longxian Village in Qingtian County, Zhejiang Province, China. The vertical distribution of this aquaculture landscape, which was constructed on a mountainside, enables downhill energy and material flows that are key to its functioning.

Rice-Fish Terraces of Longxian Village, China

Landscape Type: rice-fish terrace

Landscape Area: 60 ha (rice-fish terraces, 2019)
                      460 ha (village, 2019)

Aquaculture Yield: 1,200 kg/ha/yr (2019)

Aquaculture Type: extensive

Water Type: fresh

Rice-fish production in constructed terraces has been practiced in mountainous Qingtian County in the Zhejiang Province of China for at least 1,200 years. This long history has led to a rich tradition of rice-fish culture that informs local customs, festivals, and cuisine. Today, approximately 80% of the rice terraces in the county, almost 7,000 ha, are stocked with a red-colored variety of Common Carp (*Cyprinus carpio*), which is considered a delicacy.[1]

The area of China devoted to rice-fish culture covers approximately two million hectares in 2018, but this polyculture system is threatened by the expansion of rice and fish monoculture systems.[2] In rice–fish culture in South China, carp and rice are cultivated concurrently in one or two annual growing seasons (early rice and late rice). This cultivation system supports two methods of fish culture. High volumes of relatively small fish can be harvested multiple times annually, when fish cultivation is aligned with rice-growing seasons and harvest cycles. However, if larger carp are desired, fish can be left in the terrace to grow over the fallow winter months. In this case, fish feed on rice straw that has been left to decay.

Longxian Village is a representative rice-fish-culture village in Qingtian County. The vertical distribution of this aquaculture landscape, which was constructed on a mountainside, enables energy and material flows that are key to its functioning. Leaf litter and sediment wash down from the evergreen and deciduous forest above the terraces after rainfall, and mixes with village sewage and waste. This nutrient-rich mix is deposited in the terraces where it improves soil fertility, as opposed to polluting local rivers. Within the terraces, a mutually beneficial relationship between fish and rice exists. Carp consume insects and weeds that have a detrimental effect on rice, and their waste provide nutrients that stimulate growth. Rice, in turn, provides shade that moderates water temperature and attracts insects on which the carp feed. In rice-fish culture, no additional feed is required for fish cultivation. Studies have found that rice-fish culture requires 68% less pesticide and 24% less chemical fertilizer than rice monoculture, while producing similar rice yields.[3]

In 2005, the rice-fish culture in China was recognized as a Globally Important Agricultural Heritage System (GIAHS) by the Food and Agriculture Organization (FAO) of the United Nations. Through this initiative, the FAO has worked with local villages like Longxian to increase rice-fish production and develop and promote a culture of tourism around these ancient aquaculture landscapes. In 2006, over 10,000 tourists visited Longxian Village and the resulting increase in revenues for the village farmers provides another motivation to preserve the traditional rice-fish culture.

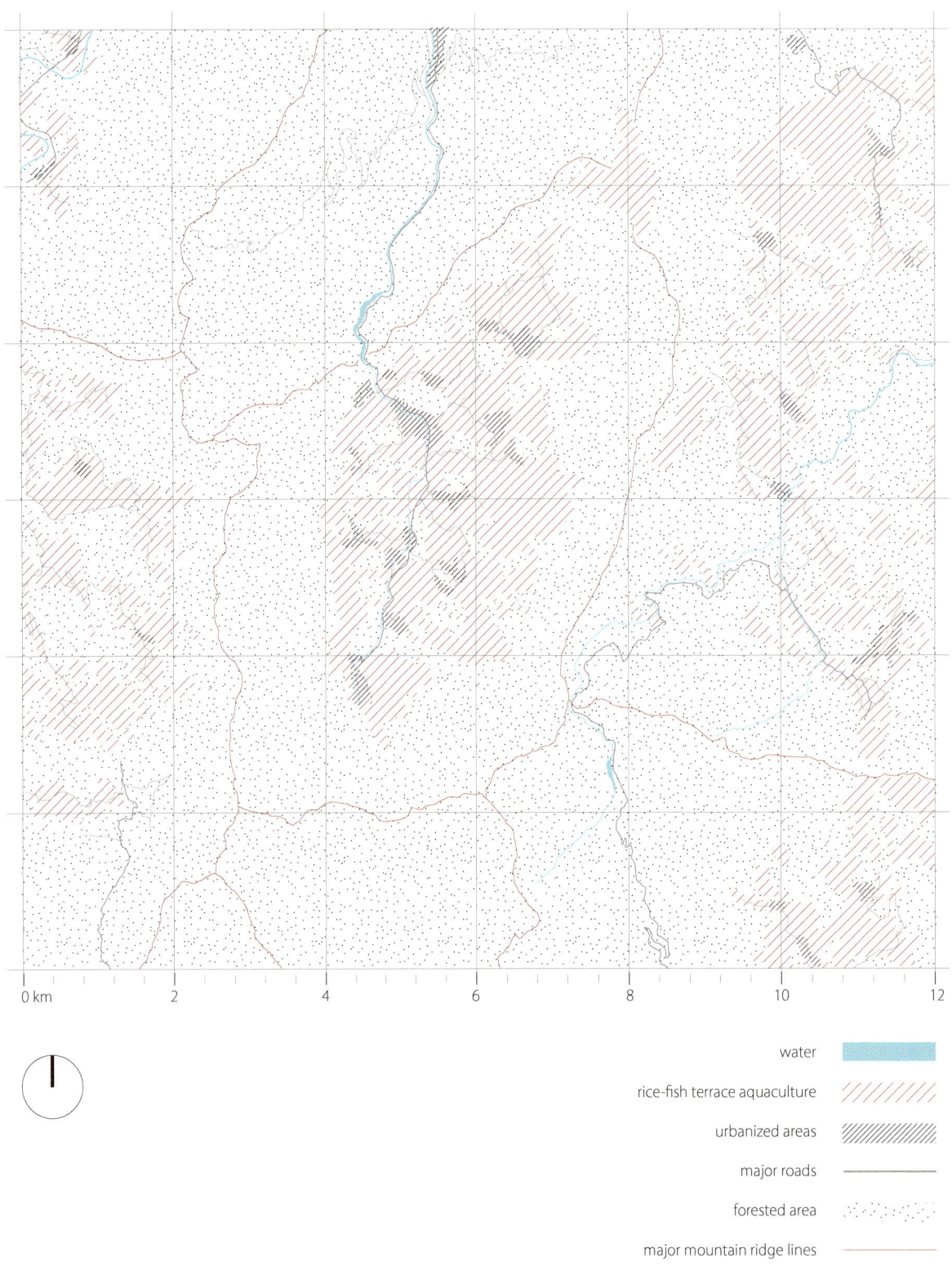

0 km   2   4   6   8   10   12

water	
rice-fish terrace aquaculture	
urbanized areas	
major roads	
forested area	
major mountain ridge lines	

02 *Regional Diagram.* The landscape of Qingtian County, China is diverse and includes coastline, hills, and mountains. Rice production in terraces constructed on the mountain slopes has been practiced in Qingtian County for over 1,200 years. In the twenty-first century, approximately 80% of the rice terraces in the county, totaling almost 7,000 ha, are stocked with carp.

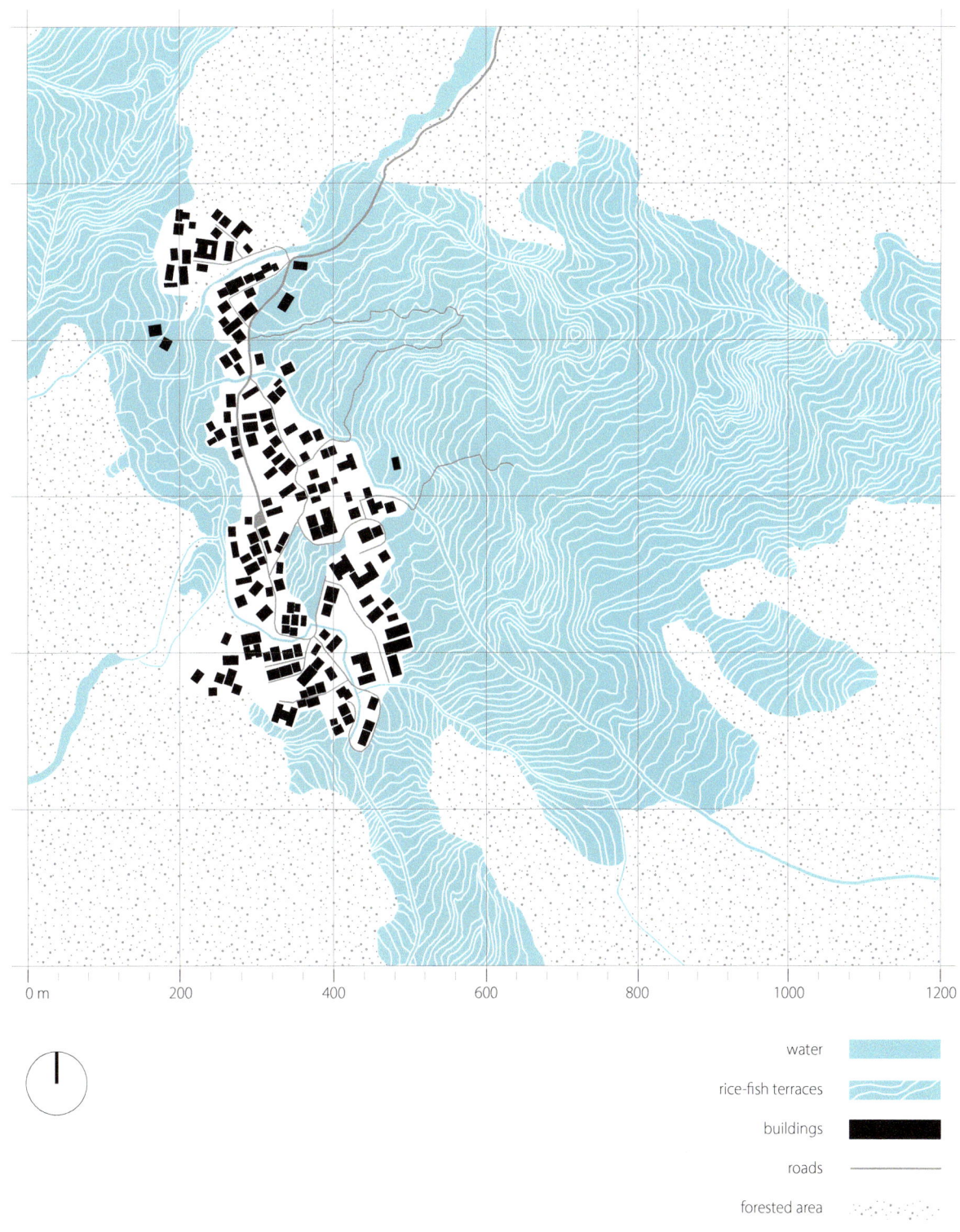

water

rice-fish terraces

buildings

roads

forested area

0 m   200   400   600   800   1000   1200

03 *Site Diagram*. Longxian Village, in the Southeast of Fangshan Town, features rice-fish terraces fed by mountain streams. The village was recognized as a Globally Important Agricultural Heritage System (GIAHS) by the Food and Agriculture Organization of the United Nations in 2005.

40

0m

0m

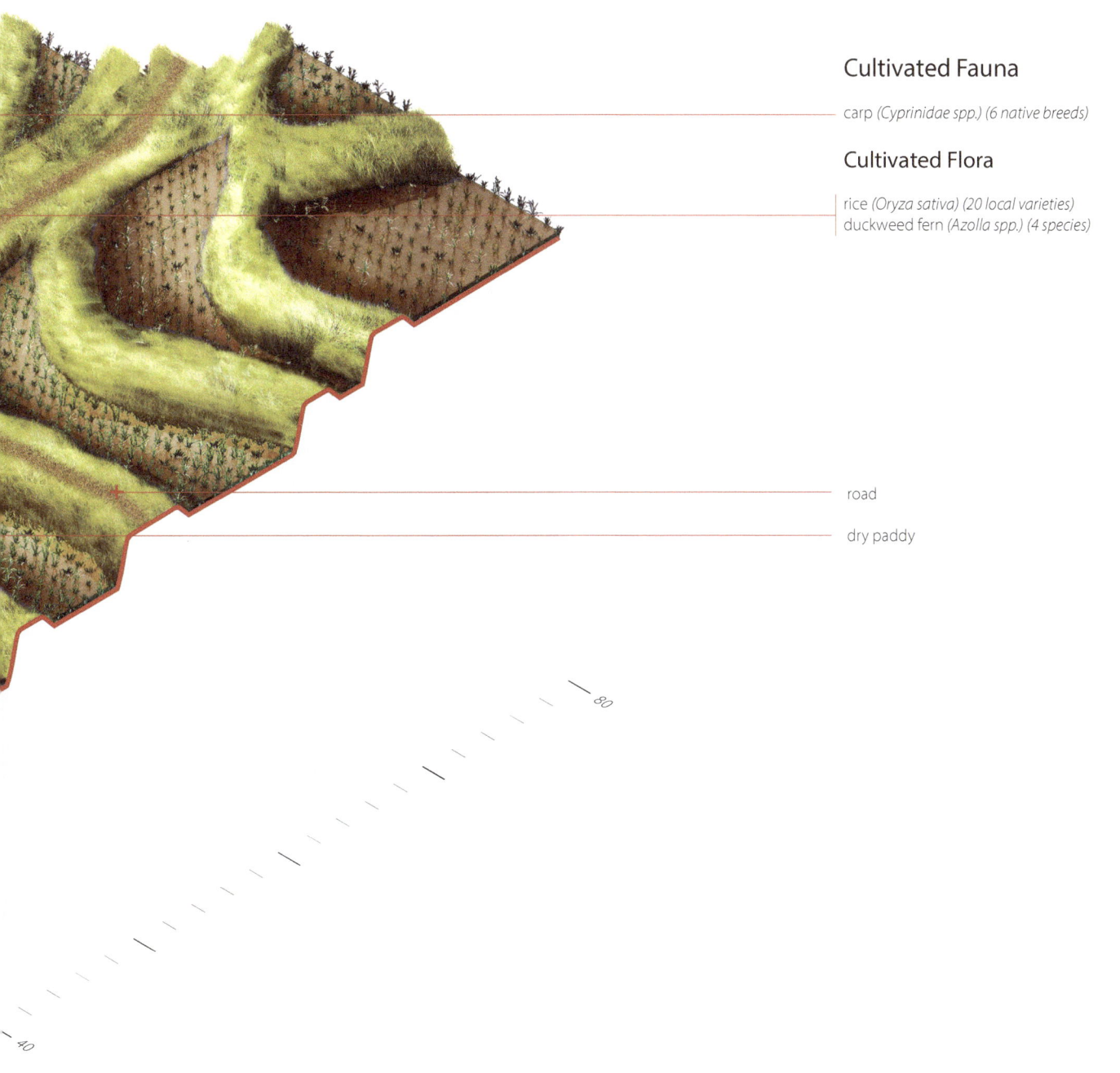

Cultivated Fauna

carp *(Cyprinidae spp.)* *(6 native breeds)*

Cultivated Flora

rice *(Oryza sativa)* *(20 local varieties)*
duckweed fern *(Azolla spp.)* *(4 species)*

road

dry paddy

80

40

04 *Transect.* Earthen dikes create terraces of varying widths that follow the contours of the mountainside at Longxian Village. A range of rice cultivars planted in the terraces creates a biodiverse mosaic. Terraces not planted with rice typically contain rice straw, a remainder from the previous rice harvest, which is left to decay and serve as food for fish.

05 View to a group of tourists walking along the earthen terrace walls that follow the contours of the mountainside.
06 Rice plants provide shade for carp in flooded terraces, and carp consume rice pests and fertilize rice with their waste. Rice-fish mutualism reduces the need for the pesticides and chemical fertilizers that are typically applied in rice monoculture.

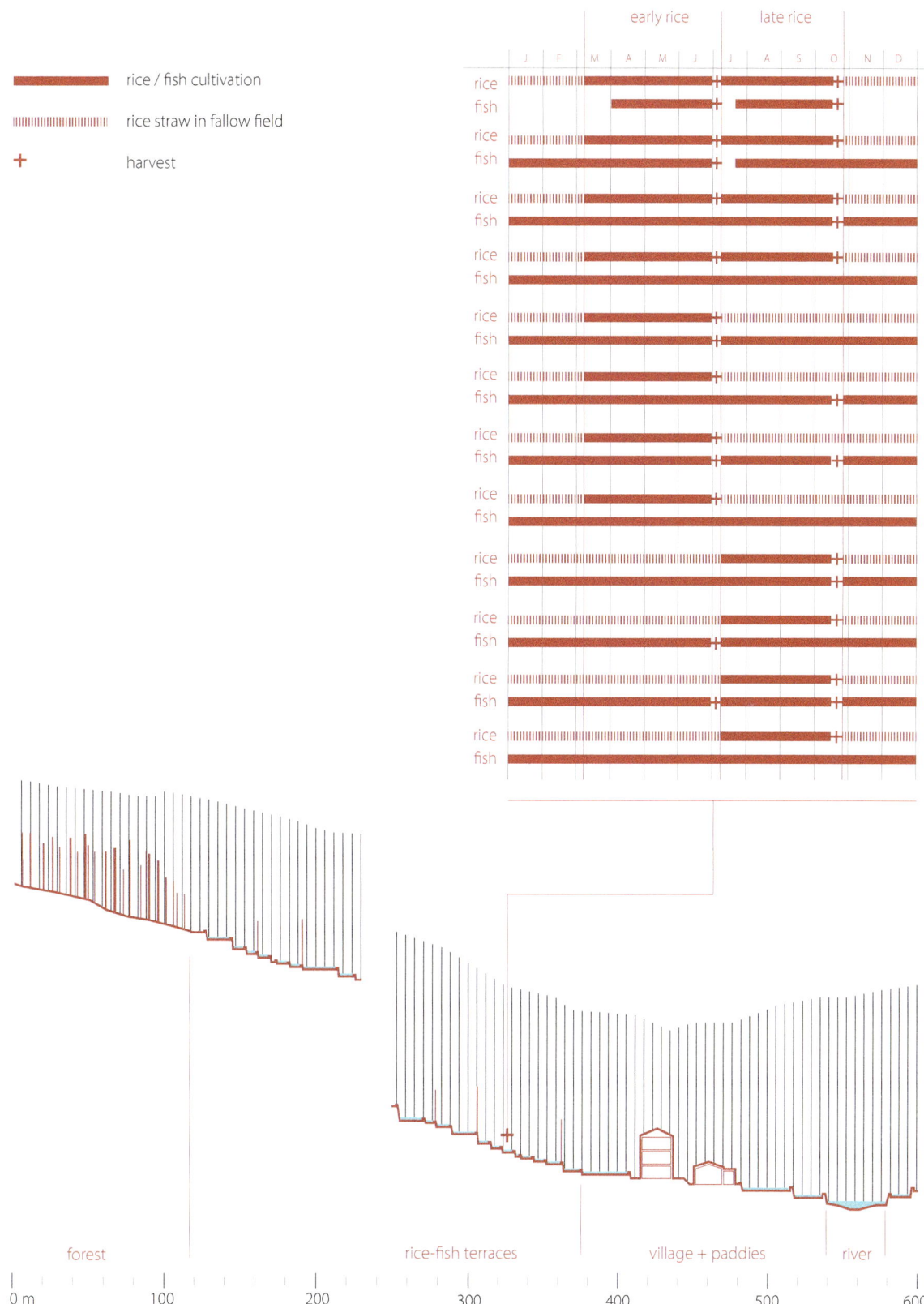

07 *Landscape Systems and Strategies.* The vertical distribution of the forest-terrace-village-paddy-river system allows for stormwater to convey nutrients from forest litter and livestock manure to the terraces via earthen ditches. At the terraces there are many permutations of rice-fish coculture, which allows flexibility in terms of crop types, yields, and harvest times.

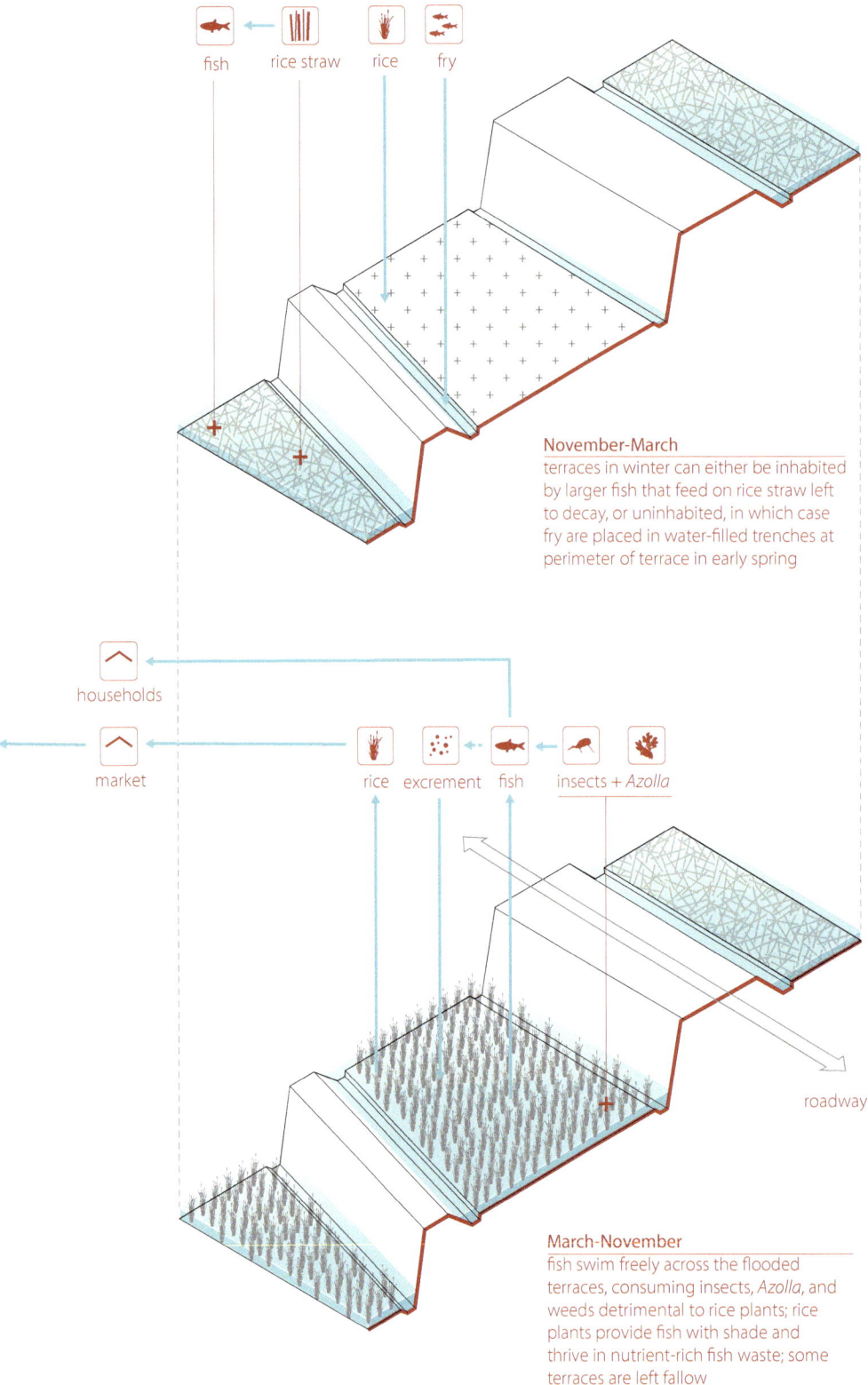

fish   rice straw   rice   fry

November-March
terraces in winter can either be inhabited
by larger fish that feed on rice straw left
to decay, or uninhabited, in which case
fry are placed in water-filled trenches at
perimeter of terrace in early spring

households

market   rice   excrement   fish   insects + *Azolla*

roadway

March-November
fish swim freely across the flooded
terraces, consuming insects, *Azolla*, and
weeds detrimental to rice plants; rice
plants provide fish with shade and
thrive in nutrient-rich fish waste; some
terraces are left fallow

08   *Landscape Systems and Strategies.* Interlaced terraces are managed for rice, fish, or rice-fish coculture, and these terraces are either flooded, drained, or semi-drained over the course of a year based on fish and rice life cycles. The range of cultivation practices across this landscape create an ever-changing mosaic of saturation levels.

09

10

11

09 Farmers harvest rice at a terrace.
10 A sluice at a dike allows for passive drainage and water flow between terraces.
11 Narrow mountainside terraces transition into broad paddies on flatter land. Forested mountains rise beyond the village.

01 Aerial view of gei wai and fishponds at the Mai Po Marshes in Hong Kong. The city of Shezhen, one of the most populous cities in China, looms directly across the Shenzhen River.

Gei Wai in the Mai Po Marshes, Hong Kong

Landscape Type: gei wai and fishpond

Landscape Area: 240 ha (gei wai, 2019)
                   1,200 ha (fishponds, 2019)

Aquaculture Yield: 60 kg/ha/yr (gei wai, pre-1990)

Aquaculture Type: extensive/ semi-intensive

Water Type: fresh/ brackish

Gei wai, traditional tidal shrimp ponds constructed in coastal areas, have a long history in Asia. Gei wai were constructed in the Mai Po Marshes in the Inner Deep Bay of Hong Kong only relatively recently, in the mid-twentieth century. Today, gei wai and commercial fishponds in this area constitute an aquaculture landscape that provides a critical habitat for migratory birds and supports ecotourism.

Gei wai in the Mai Po Marshes were constructed by dredging water channels around stands of mangroves, then using the dredged mud to build dikes to enclose the water and mangroves. Opening a sluice gate on the bayside dike during seasonal high tide flushed young shrimp into the gei wai from Deep Bay. The shrimp were then harvested as water flowed out at low tide. Farmers managed the mangrove stands inside the pond as a source of food as well as shade for the shrimp.[1]

Beginning in the 1960s, shrimp farmers began modifying hundreds of hectares of gei wai in the Mai Po Marshes to create more lucrative deep-water fishponds for carp polyculture. In the twenty-first century, only twenty-one gei wai, covering 240 ha, remain within the Mai Po Marshes. The commercial fishponds in this area cover over 1,000 ha and they serve as the major supplier of freshwater fish to the New Territories of Hong Kong. In 2005, a 1,500 ha area of the Mai Po Marshes was designated a RAMSAR Wetland of International Importance.

Wen Xianji, assistant director of the Mai Po reserve, describes the relationship between shrimp, humans, and birds at the site, observing that "when we harvest there are small fish and shrimp we don't want, and the birds come and eat. This traditional way of aquaculture benefits both humans and birds."[2] Adjustments made to the water levels in the gei wai and fishponds throughout the year are key to this symbiotic relationship between species. In winter, many gei wai and fishponds are drained of water to harvest the remaining shrimp and fish. This drawdown exposes fish, worms, and amphibians and creates an important feeding opportunity for thousands of water birds.[3] In other seasons water levels at the gei wai that are not used for shrimp production are raised and lowered to create optimal nesting and roosting habitats for migratory birds.

Today, the World Wildlife Fund (WWF) operates only a small handful of gei wai for shrimp harvest for educational purposes. Tourists join the traditional nighttime shrimp harvests with flashlights and help to sort the captured shrimp. Other infrastructure for walkers, cyclists, and birders at the gei wai and larger Mai Po Marshes Nature Reserve include a series of bird blinds, extensive trails and boardwalks, and an education center that is managed by the WWF.[4]

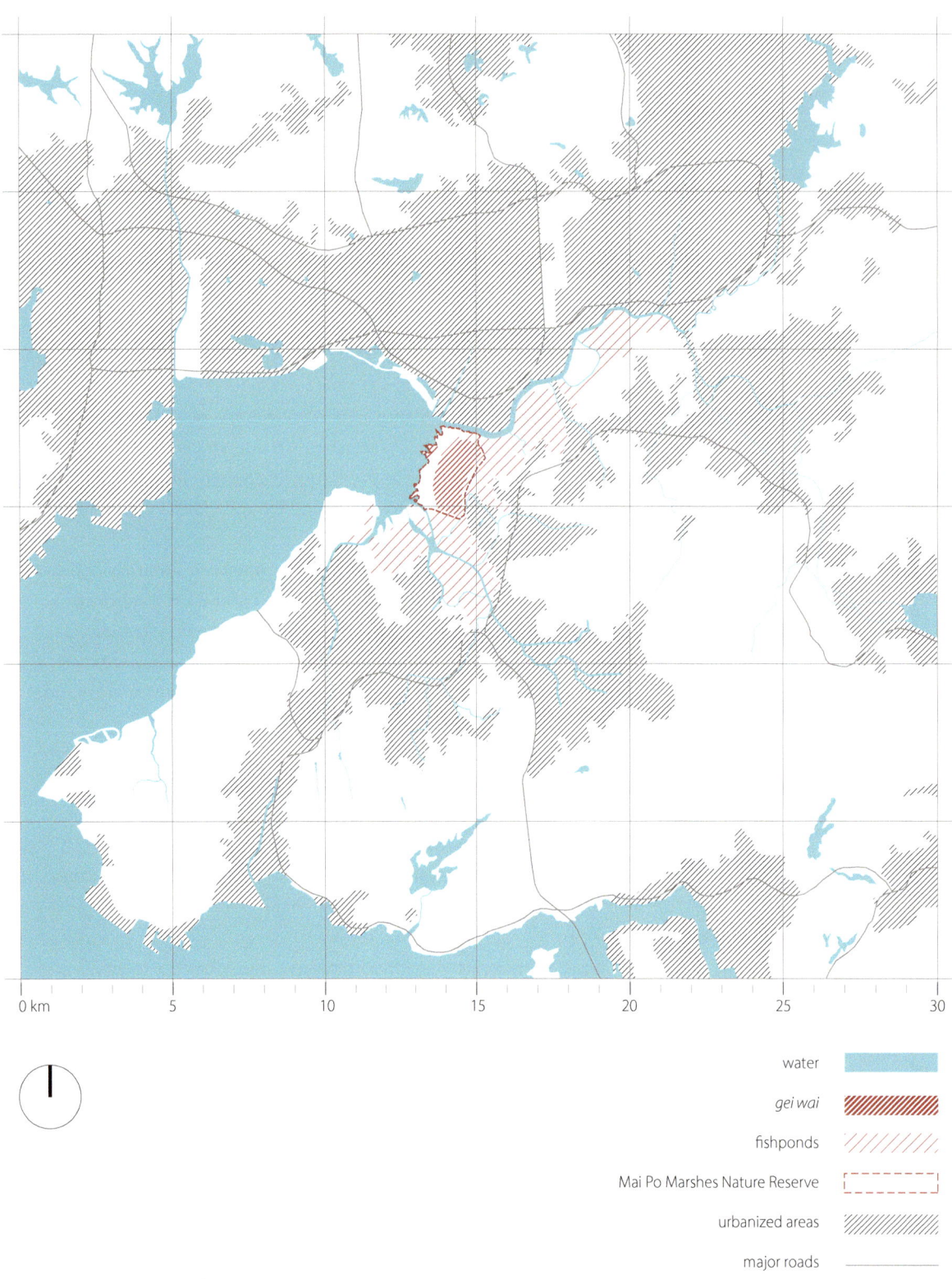

water	
gei wai	
fishponds	
Mai Po Marshes Nature Reserve	
urbanized areas	
major roads	

02 *Regional Diagram.* Aquaculture landscapes in northwest Hong Kong line the Inner Deep Bay and the southern edge of the Shenzhen River. In the mid-twentieth century, gei wai were the predominant landscape type in this area. In the twenty-first century, gei wai are found only within the Mai Po Marshes Nature Reserve and cover 300 ha.

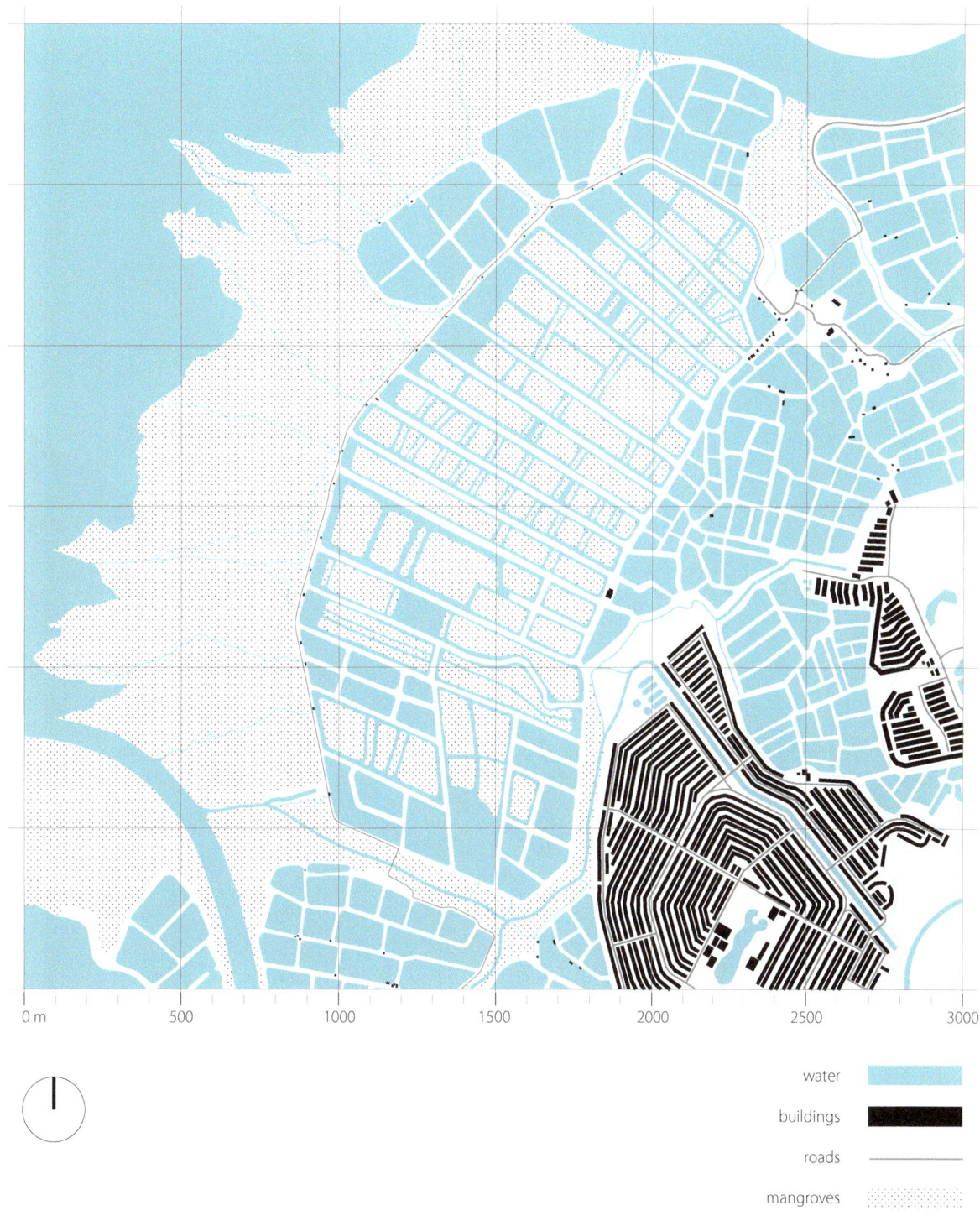

| 0 m | 500 | 1000 | 1500 | 2000 | 2500 | 3000 |

water

buildings

roads

mangroves

03 *Site Diagram.* The World Wildlife Fund (WWF) manages the Mai Po Marshes and only three gei wai in the reserve are operated using traditional practices. Other gei wai are managed as critical habitat for piscivorous birds. The fishponds surrounding the reserve also serve an ecological function by providing habitat for wintering waterfowl.

Cultivated Fauna

Gei Wai Shrimp *(Metapenaeus ensis)*
Flathead Grey Mullet *(Mugil cephalus)*
Bighead Carp *(Aristichthys nobilis)*
Silver Carp *(Hypophthalmichthys molitrix)*
Grass Carp *(Ctenopharyngodon idella)*
Common Carp *(Cyprinus carpio)*
Nile Tilapia *(Oreochromis nilotica)*

Cultivated Flora

mangroves *(Kandelia candel)*

80

fisherman's hut

shrimp-capture sluice

40

dirt access road

waterway through mangrove forest

04 *Transect*. A characteristic feature of a gei wai is the mangrove island encircled by 2 m deep water channels. The mangroves provide shade and food for shrimp and fish in the channels. The stone-core dike between the Deep Bay and gei wai is penetrated by a concrete sluice that provides control over inflow and outflow of brackish water in the system.

05 View to a gei wai from a bird blind. This gei wai is managed as biodiverse habitat rather than for shrimp harvest.
06 Satellite image in 2019 of the gei wai maintained by the WWF as habitat for the endangered Black-faced Spoonbill. The water level here is lowered in winter to provide roosting habitat and raised in summer to manage invasive plant species.

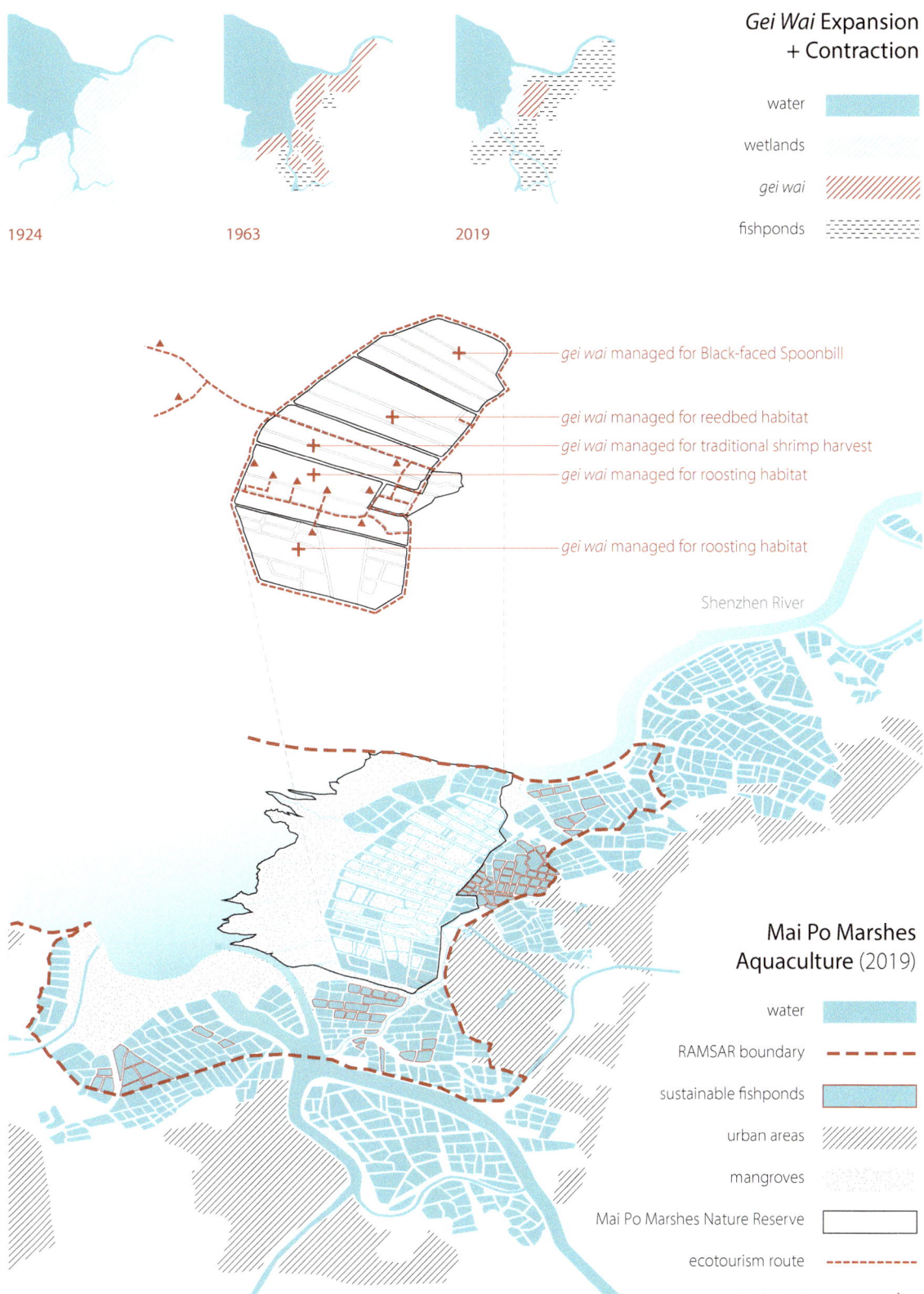

Gei Wai Expansion
+ Contraction

water

wetlands

gei wai

fishponds

1924   1963   2019

gei wai managed for Black-faced Spoonbill

gei wai managed for reedbed habitat
gei wai managed for traditional shrimp harvest
gei wai managed for roosting habitat

gei wai managed for roosting habitat

Shenzhen River

Mai Po Marshes
Aquaculture (2019)

water

RAMSAR boundary

sustainable fishponds

urban areas

mangroves

Mai Po Marshes Nature Reserve

ecotourism route

bird blind

07 *Landscape Systems and Strategies.* The number of gei wai in the region has been in decline since around the mid-twentieth century; most were converted into conventional fishponds. Today the WWF operates the remaining gei wai and has constructed infrastructure for ecotourism that includes boardwalks, bird blinds, and an education center.

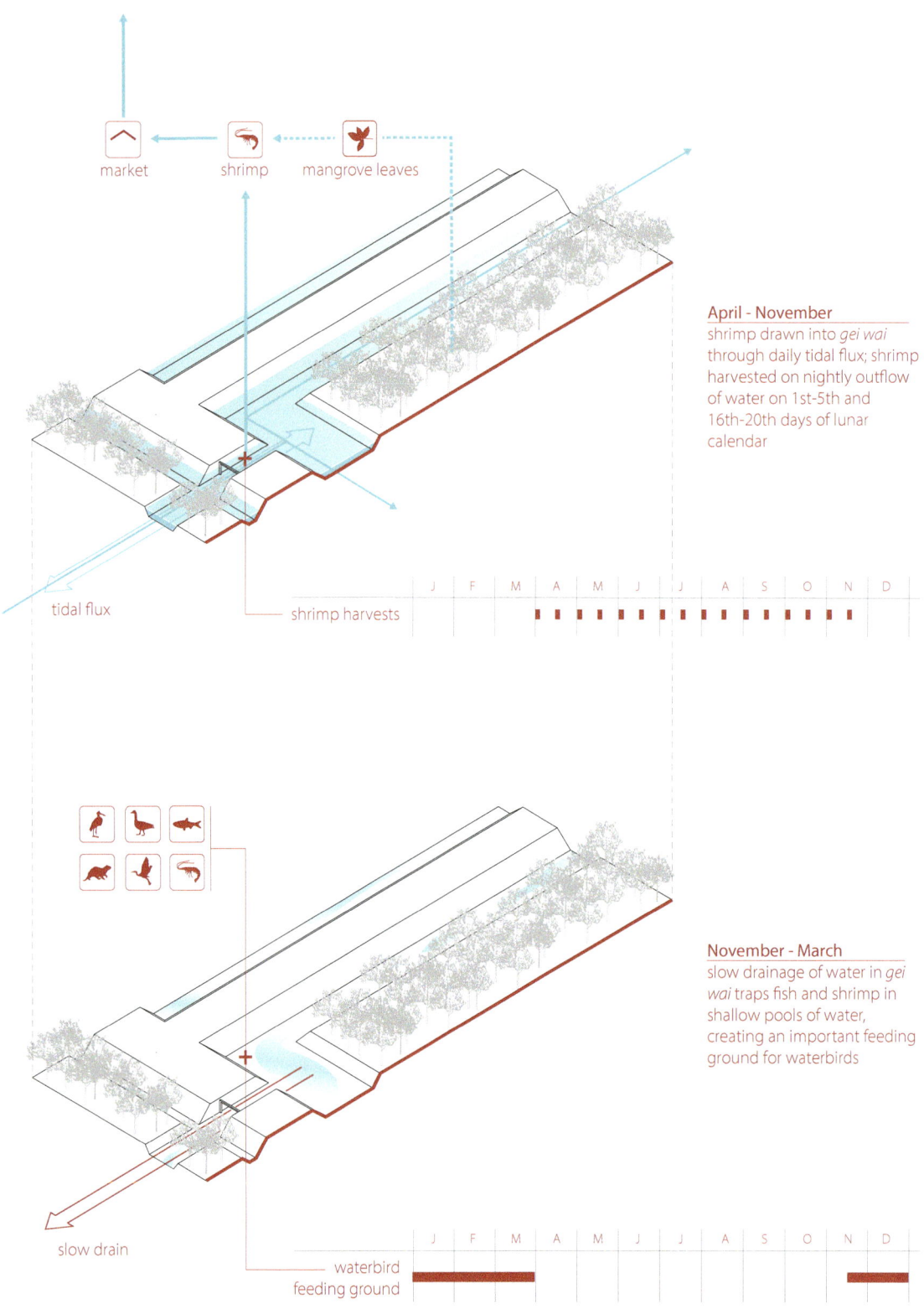

market   shrimp   mangrove leaves

April - November
shrimp drawn into *gei wai* through daily tidal flux; shrimp harvested on nightly outflow of water on 1st-5th and 16th-20th days of lunar calendar

J F M A M J J A S O N D

tidal flux

shrimp harvests

November - March
slow drainage of water in *gei wai* traps fish and shrimp in shallow pools of water, creating an important feeding ground for waterbirds

J F M A M J J A S O N D

slow drain

waterbird feeding ground

08 *Landscape Systems and Strategies.* Traditional operation of a gei wai is based on the capture of shrimp that flow into and out of the gei wai with tidal flux. At the Mai Po Marshes only three gei wai are managed for shrimp harvest, the water levels at the 18 other gei wai are managed to create optimal nesting, roosting and feeding grounds for migratory birds.

09

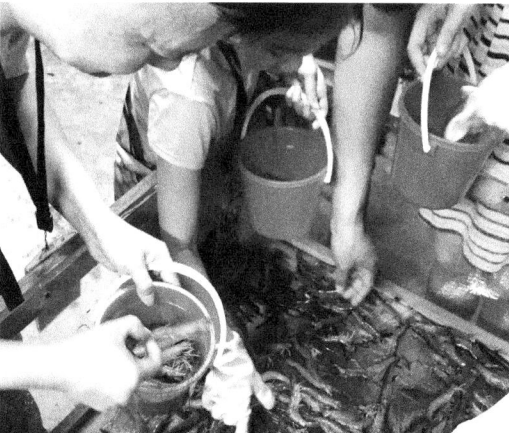

10
11

09 Egrets feeding on remaining fish and shrimp following a winter gei wai water drawdown.
10 Operation of a sluice during a nighttime shrimp harvest demonstration at a gei wai managed for educational purposes.
11 Visitors to a nighttime shrimp harvest demonstration assist in sorting the captured shrimp.

01 Satellite image illustrates the zig-zag form of the pre-Columbian savanna weirs that span the savanna between forest islands in the Baures region of Bolivia.

Savanna Weirs of Baures, Bolivia

Landscape Type: savanna weir

Landscape Area: 52,000 ha (~16th C)

Aquaculture Yield: 1,000 kg/ha/yr (~16th C) | 600 kg/ha/yr (2019)

Aquaculture Type: extensive

Water Type: fresh

The Baures region of the Bolivian Amazon is characterized by grassland savannas, forest islands, and wetlands. In this region, a range of pre-Columbian earthworks were constructed by the people of Baure prior to European occupation in the early eighteenth century. These works include canal-causeways, raised agricultural fields, and settlement mounds. Anthropologist Clark Erickson, a leading scholar of pre-Columbian cultural landscapes, theorizes that earthen weirs are yet another of these earthworks, and the weirs transformed the savanna into a productive, aquaculture landscape when it flooded annually. Erickson writes that by "using this simple, but elegant, technology, the people of Baure converted much of the landscape into an aquatic farm . . . rather than domesticate the species that they exploited, the people of Baure domesticated the landscape."[1]

During the wet season, fish migrate to and spawn in inundated savannas of the Baures region. It is thought that low, jagged dikes were constructed across the savanna, both to impound shallow waters and to create the means to harvest fish. These savanna weirs have frequent changes in direction, and many extend distances of up to 3.5 km. Where the weirs form a sharp angle, funnel-like openings are present, suggesting that woven baskets may have been placed within these openings to capture fish moving through them. These extant weirs cover over 500 km^2 of the savanna. A total of 1,515 kilometers of weirs are estimated to have been constructed in this region.[2]

Across the savanna, circular ponds are associated with the weirs, with some as big as 30 m in diameter. In the past, these reservoirs likely functioned to hold water throughout the year. They also may have been utilized by the people of Baure to store live fish during the dry season. Even today, the extant reservoirs continue to fill passively with fish after seasonal floods. A palm tree, known locally as Carandai-guazú (*Mauritia flexuosa*), may have been cultivated on the fish weirs and reservoirs. The tree was a food source that produced edible fruit and starch, and the trunk and fronds were harvested by the people of Baure to produce basketry, mats, and roof beams.[3]

The people of Baure lived in dozens of large settlements that were located on intermittent forest islands within the savanna.[4] The Baures savanna is unpopulated today but archaeological work exploring the historical earthworks is ongoing. It is thought that the savanna weirs, coupled with Carandai-guazú, provided food and more for a regional population of 40,000 people. In Erickson's evocative characterization, the flooded savanna was "a sea of hundreds of square kilometers of starch and protein."[5] As it did in the past, this constructed landscape continues to have a profound effect on the hydrology and biodiversity of the region.[6]

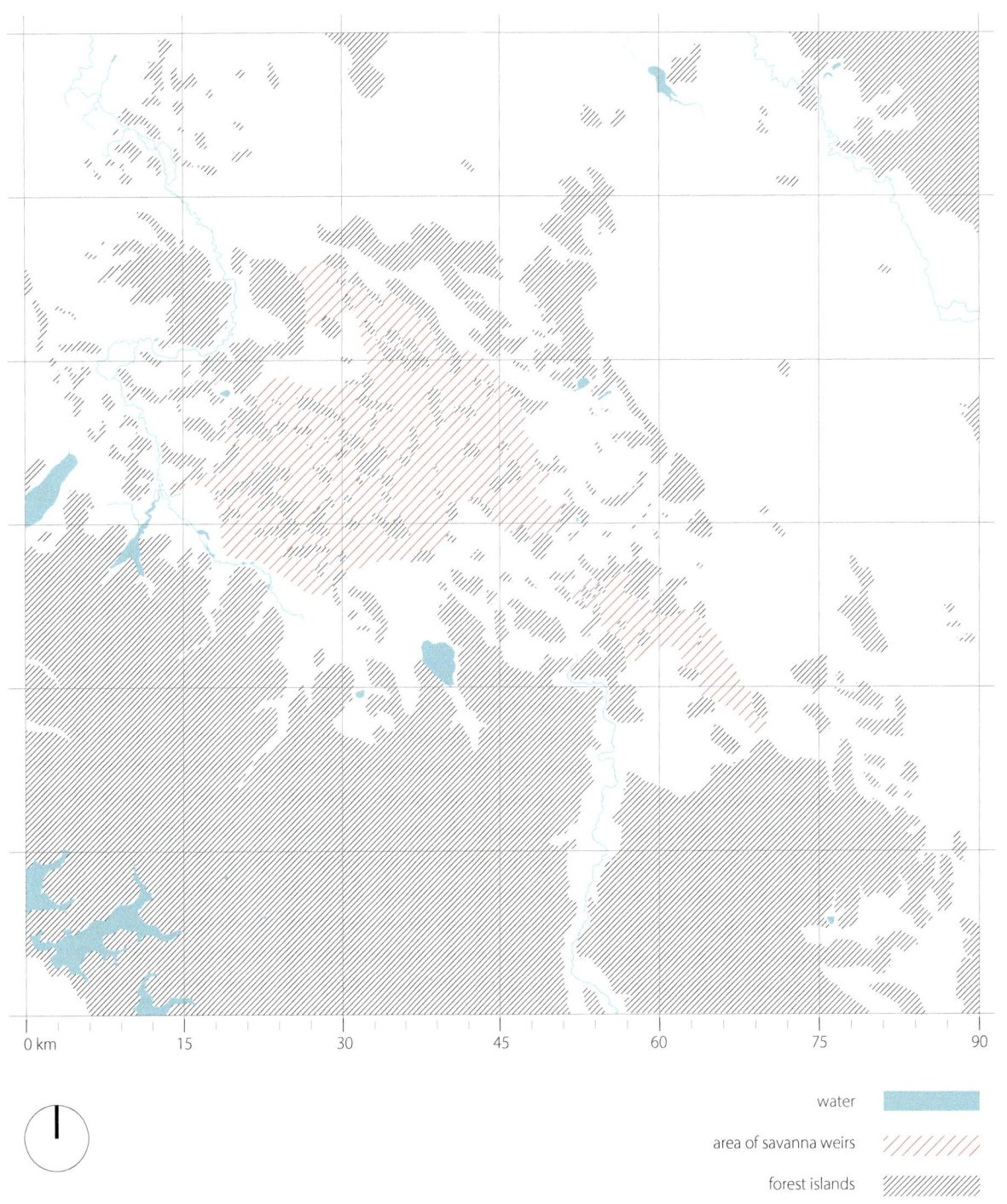

0 km   15   30   45   60   75   90

water

area of savanna weirs

forest islands

02 *Regional Diagram.* The Baures region is characterized by forest islands within a seasonally flooded grassland savanna. The people of Baure constructed weirs and a range of other earthworks in this landscape. This diagram reproduces a map originally published in the article "An Artificial Landscape-scale Fishery in the Bolivian Amazon," by Clark Erickson.

0 m   200   400   600   800   1000   1200

water at flooded savanna	
weirs	
canal-causeways	
reservoirs	
forest island	
scattered tree patches	

03 *Site Diagram.* Forest islands in the Baures region were interconnected with two types of earthworks. Canal-causeways were straight, elevated road and waterways used for transportation and water management, and savanna weirs were jagged dikes thought to have been used to manage and harvest fish in the flooded savanna. After a map by Clark Erickson.

Cultivated Fauna

buchere *(Hoplosternum spp.)*
Cunare *(Cichla monoculos)*
palometa *(Serrasalmus spp.)*
Sabalo *(Prochilodus nigricans)*
benton *(Erythrina spp.)*
Spotted Applesnail *(Pomacea maculata)*

Cultivated Flora

carandai-guazú *(Mauritia flexuosa)*

circular reservoir for live fish storage

earthen berm

flooded savanna

fish weir

fish capture basket

40

0m

04 *Transect.* Baskets at breaks in the earthen savanna weirs captured fish swimming across the flooded savanna during the wet season. The weirs were 1 m to 2 m wide and approximately 0.5 m high. Circular reservoirs measured 10 m to 30 m in diameter; the largest reservoirs likely held water throughout the year and may have been used for live fish storage.

05 Aerial view of circular reservoirs and savanna weirs spanning between forest islands.
06 Intermittent V-shaped breaks in the savanna weirs acted as funnels that directed the flow of water across the savanna. They also served as productive sites for fish capture at times of the year when the savanna was flooded.

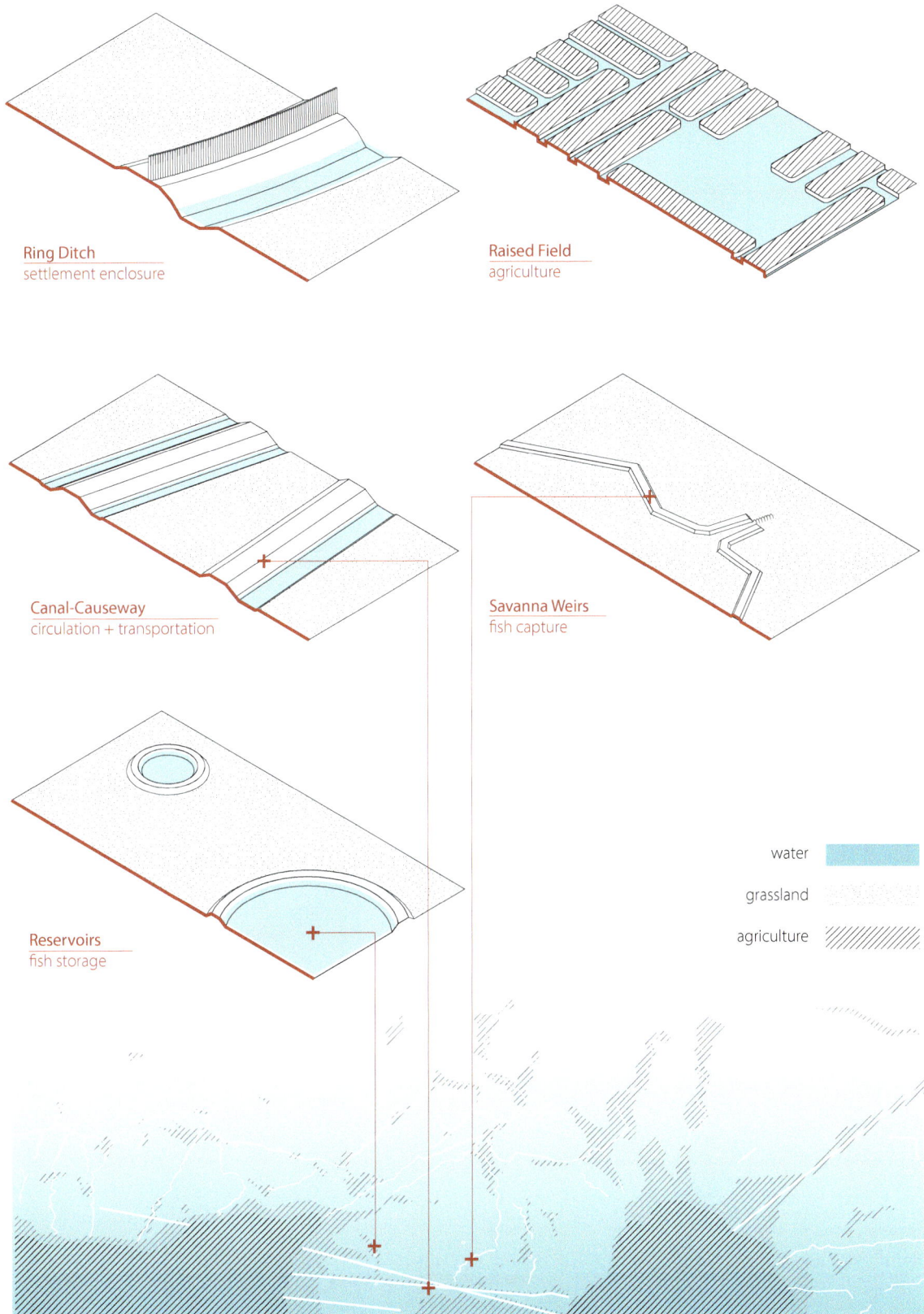

Ring Ditch
settlement enclosure

Raised Field
agriculture

Canal-Causeway
circulation + transportation

Savanna Weirs
fish capture

Reservoirs
fish storage

water
grassland
agriculture

07 *Landscape Systems and Strategies.* The savanna weirs and circular reservoirs constructed for aquaculture joined other earthworks built to facilitate transportation, manage water, structure settlement boundaries, and enable agricultural production.

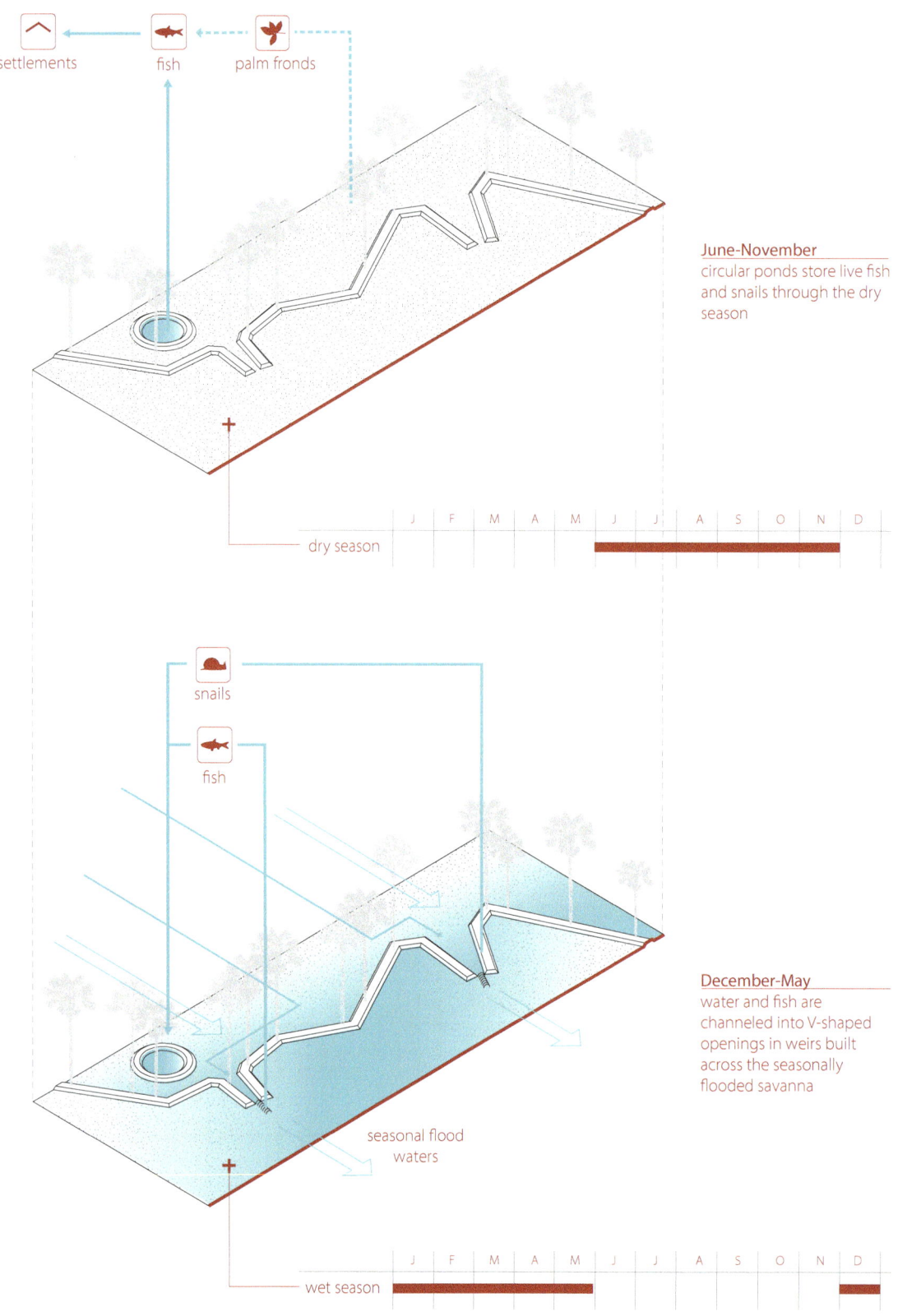

settlements   fish   palm fronds

June-November
circular ponds store live fish
and snails through the dry
season

J	F	M	A	M	J	J	A	S	O	N	D

dry season

snails

fish

December-May
water and fish are
channeled into V-shaped
openings in weirs built
across the seasonally
flooded savanna

seasonal flood
waters

J	F	M	A	M	J	J	A	S	O	N	D

wet season

08 *Landscape Systems and Strategies.* The savanna weirs and reservoirs produced abundant and storable yields of protein.
The weirs contain seasonally flooded areas of 10 ha to 80 ha. The V-shaped breaks in the weirs occur every 10 m to 30 m.
Large numbers of Spotted Applesnail (*Pomacea maculata*) associated with the weirs may have been cultivated as food.

09 Aerial view of savanna weirs spanning between forest islands. Today shrubs, trees and termite mounds cover the weirs.
10 View to a stand of palm trees, locally known as Carandai-guazú *(Mauritia flexuosa)* in the savanna. The tree is thought to have been cultivated on the weirs by the people of Baure.

01 Aerial view of Hatchery Creek, the Wolf Creek National Fish Hatchery, and the Wolf Creek Dam at the Cumberland River. These three interrelated environments for fish production and angling constitute a unique aquaculture landscape within the US National Fish Hatchery System.

Tailwater Fisheries at Wolf Creek Dam, United States

Landscape Type:	tailwater fisheries and hatchery
Landscape Area:	130 km (tailwater) | 2,000 ha (hatchery) 1.8 km (trout stream)
Aquaculture Yield:	125,000 kg/yr (hatchery)
Aquaculture Type:	intensive
Water Type:	fresh

Construction of the Wolf Creek Dam in Kentucky in the mid-twentieth century held back the water of the Cumberland River and created Cumberland Lake. Today, water drawn from the depths of the lake supplies three distinct elements of an aquaculture and angling landscape adjacent to the dam—the Cumberland River, the Wolf Creek National Fish Hatchery, and Hatchery Creek.

Downstream of the dam the Cumberland River is a tailwater, a reservoir-fed river that is characterized by consistently low water temperatures, moderate flow, clarity, and nutrient abundance. These qualities create an optimized trout habitat that is prized by recreational anglers.[1] Fish spawning is not possible in the Cumberland River as the cyclical releases of water required by hydroelectric production at the Wolf Creek dam washes out fish eggs. Instead, each year the river is stocked with approximately 100,000 Rainbow, Brook, and Brown Trout that are raised at the Wolf Creek National Fish Hatchery, a facility located at the foot of the dam.[2] The hatchery imports and incubates fish eggs from four National Hatcheries. Juvenile trout grow out in a series of outdoor concrete raceways that are fed by the cold water of Lake Cumberland. Water from the hatchery is discharged into the newest element of this aquaculture and angling landscape—a 1,800 m constructed and self-sustaining spawning trout stream called Hatchery Creek.

Stream restoration specialist George Athanasakes, of the international design firm Stantec, convened a multidisciplinary team to design Hatchery Creek that included engineers, biologists, ecologists, vegetation specialists, anglers, and contractors. To initiate the design phase, the team "drew the life stages of trout on a board and discussed the optimum stream types and habitats needed for each life stage."[3] Completed in 2016, the stream is designed to provide habitat for trout as well as support recreational angling. Hatchery Creek has the appearance of a natural stream and features an abundance of strategically placed cover for fish. The stream features Rosgen Type A, Type C, and Type DA classifications as well as off-channel wetlands and vernal ponds. Spawning trout enter the stream from the Cumberland River via a series of step pools and lay eggs in gravel riffles. Fry occupy off-channel streams with shallow, warm water, while juveniles and adults find cover among boulders, downed timber, and cut banks in the main channel.[4]

These three elements—river, hatchery, and stream—are all constructed waterways and ecologies that draw water from Lake Cumberland, and so in some sense all three are tailwaters. While these waterways vary radically in terms of material, form, scale, and character, together they constitute a cohesive and interconnected aquaculture landscape for fish and humans in Kentucky.

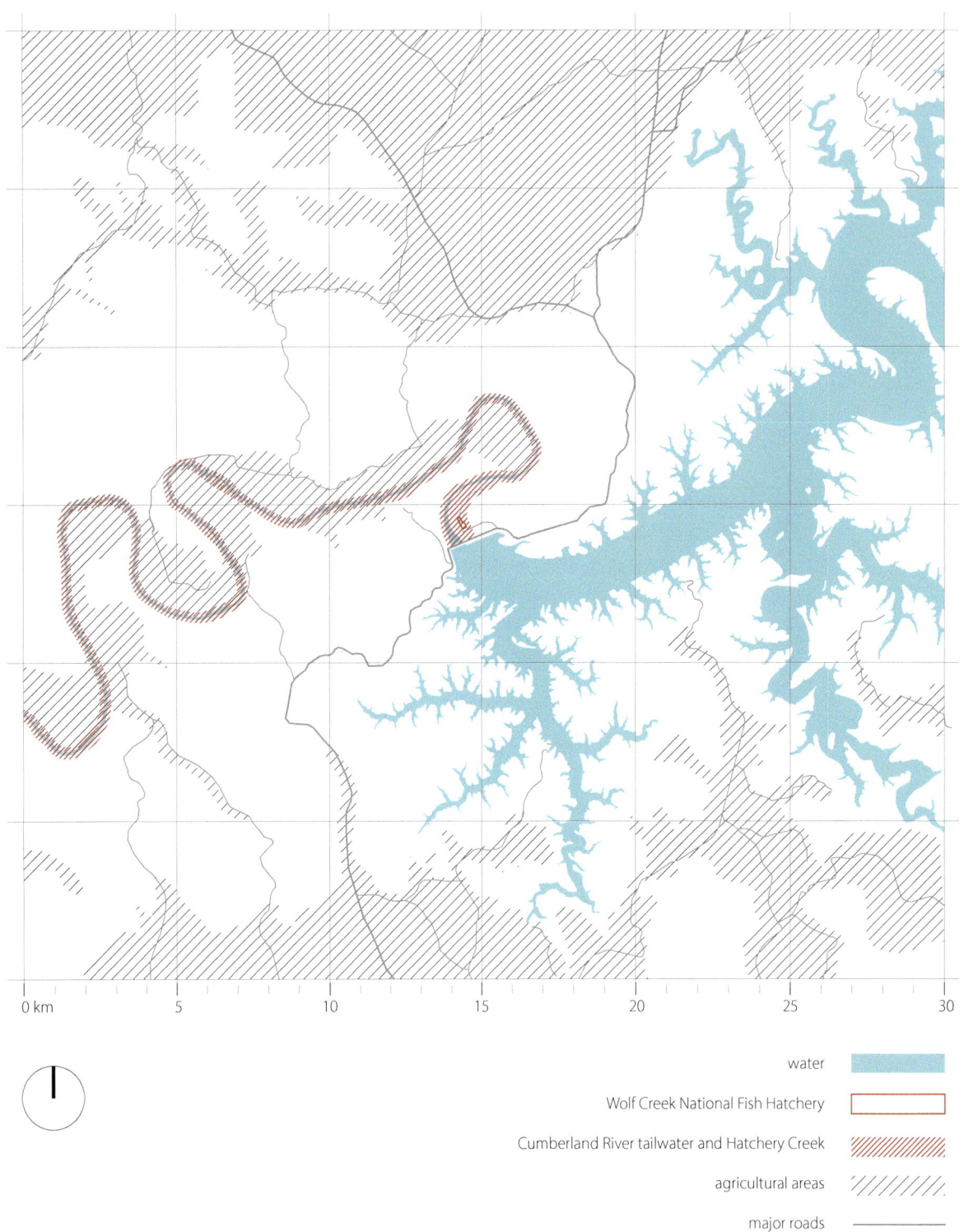

water ▬▬▬

Wolf Creek National Fish Hatchery ▭

Cumberland River tailwater and Hatchery Creek ▨

agricultural areas ▨

major roads ▬▬▬

0 km 5 10 15 20 25 30

02 *Regional Diagram*. In 1951 the Cumberland River was dammed in order to mitigate flood damage from storms, creating Lake Cumberland and a tailwater. The Wolf Creek National Fish Hatchery, located at the foot of the dam, stocks the Cumberland River as well as other regional rivers with hundreds of thousands of Rainbow, Brook, and Brown Trout each year.

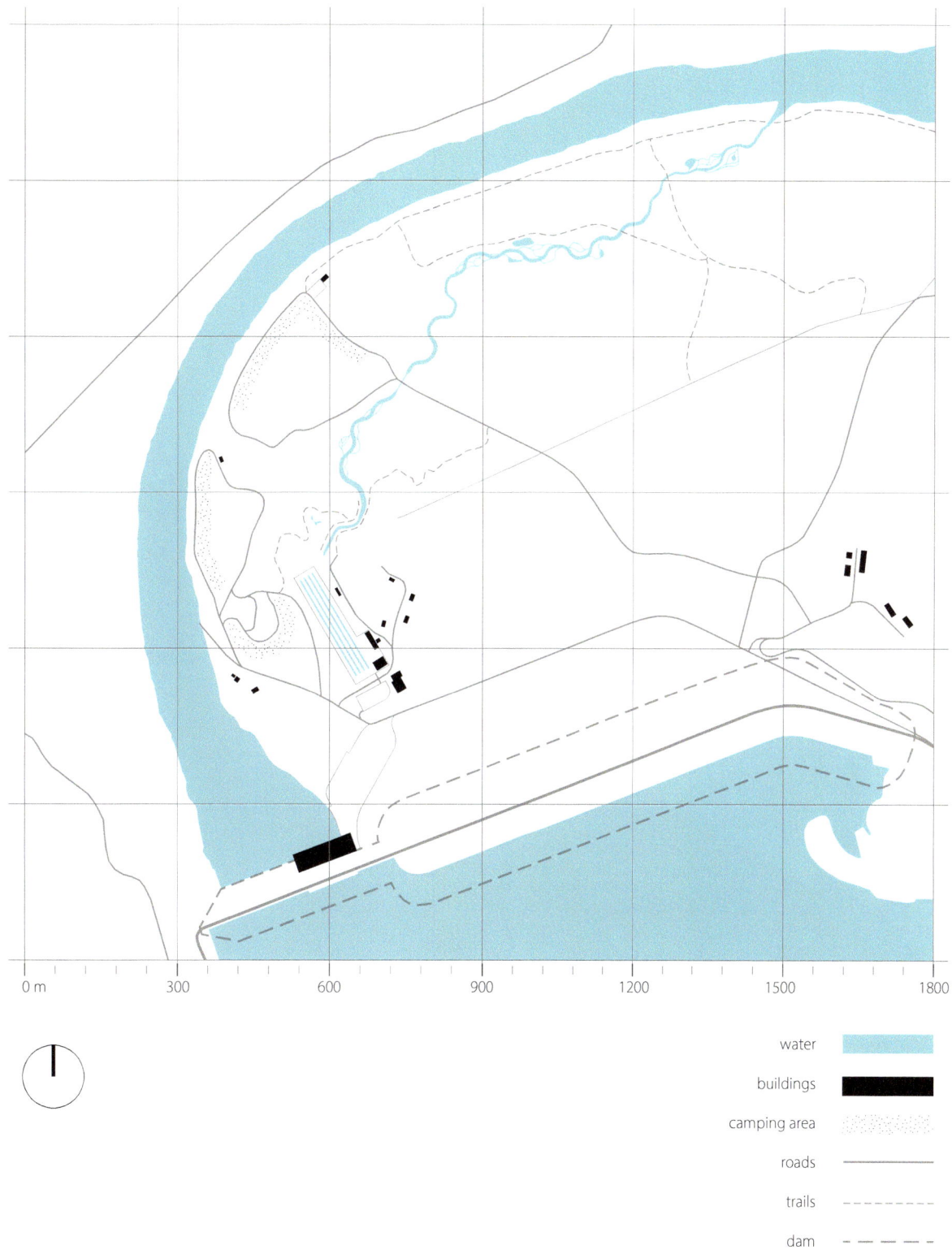

water	
buildings	
camping area	
roads	
trails	
dam	

0 m 300 600 900 1200 1500 1800

03 *Site Diagram.* Discharge from Wolf Creek National Fish Hatchery flows in the newly constructed Hatchery Creek, an 1,800 m self-sustaining trout stream. The linear raceways of the hatchery contrast with the meandering and braided channels of Hatchery Creek, where trout reproduction through spawning occurs without direct human intervention.

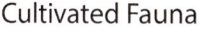

Cultivated Fauna

Rainbow Trout *(Oncorhynchus mykiss)*
Brook Trout *(Salvelinus fontinalis)*
Brown trout *(Salmo trutta)*
Tuxedo Darter *(Etheostoma lemniscatum)*
Barrens Topminnow *(Fundulus julisia)*
Relict Darter *(Etheostoma chienense)*
Logperch *(Percina caprodes)*
Spotfin Chub *(Cyprinella monacha)*

Cultivated Flora

pignut hickory *(Carya glabra)*
river birch *(Betula nigra)*
fox sedge *(Carex vulpinoidea)*
riverbank wild rye *(Elymus riparius)*

bird netting at raceways

raceway at Wolf Creek National Fish Hatchery

drained raceway

hatchery water discharge

Hatchery Creek

rock riffle

accessible angling platform

log and rock riffle

04 *Transect*. Discharge from the concrete raceways of the hatchery flows into Hatchery Creek at 60 m³/min. The portion of Hatchery Creek depicted here features constructed rock and log riffles and is stocked with trout grown at the hatchery; a migration barrier downstream prevents trout that spawned in Hatchery Creek from mixing with hatchery-reared trout.

05 View to rock riffle placed at Type C stream in Hatchery Creek.
06 The slow water in a wetland zone designed among braided channels at Hatchery Creek creates habitat ideal for trout in early stages of their life cycle.

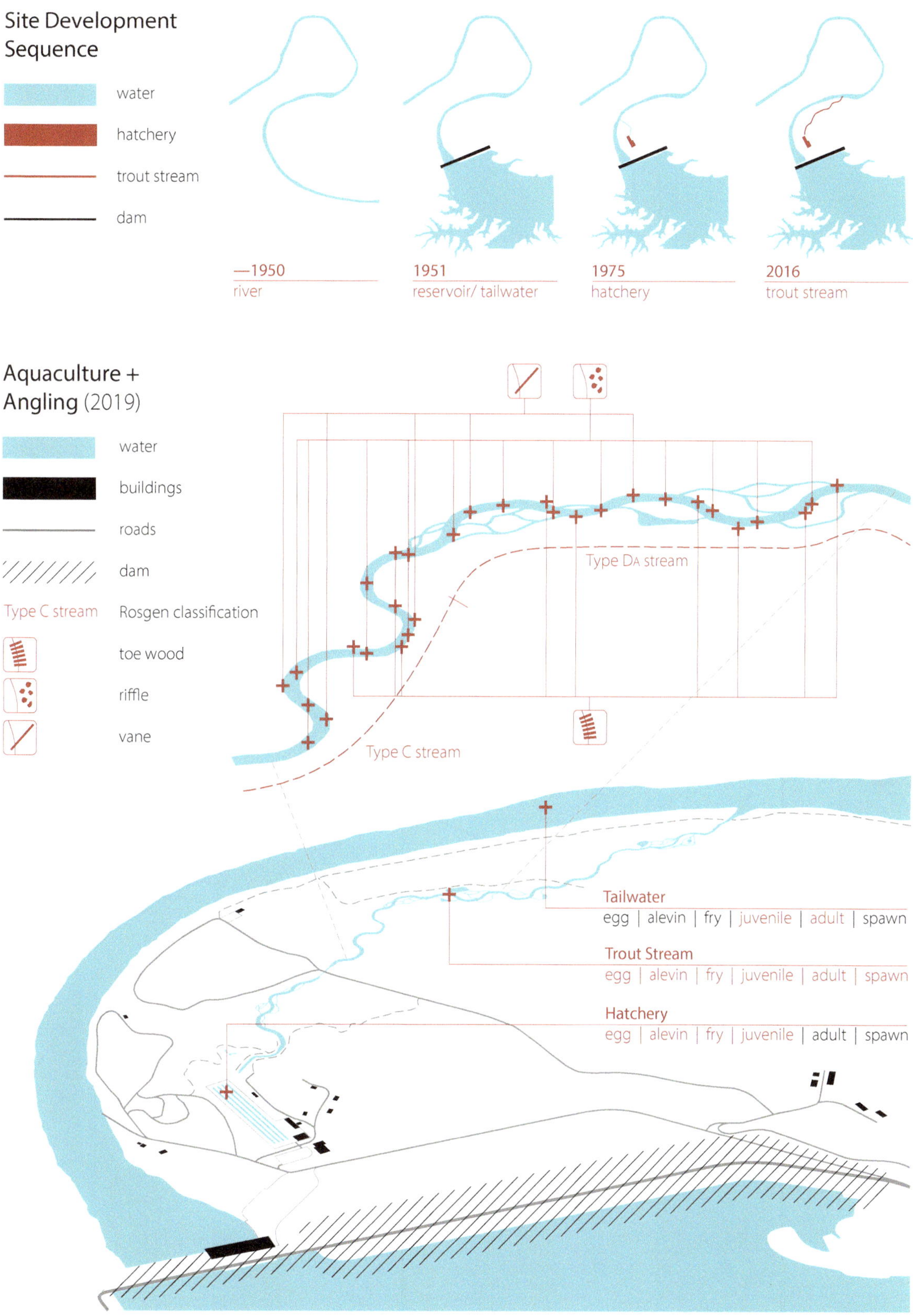

Site Development Sequence

- � water
- ▬ hatchery
- ▬ trout stream
- ▬ dam

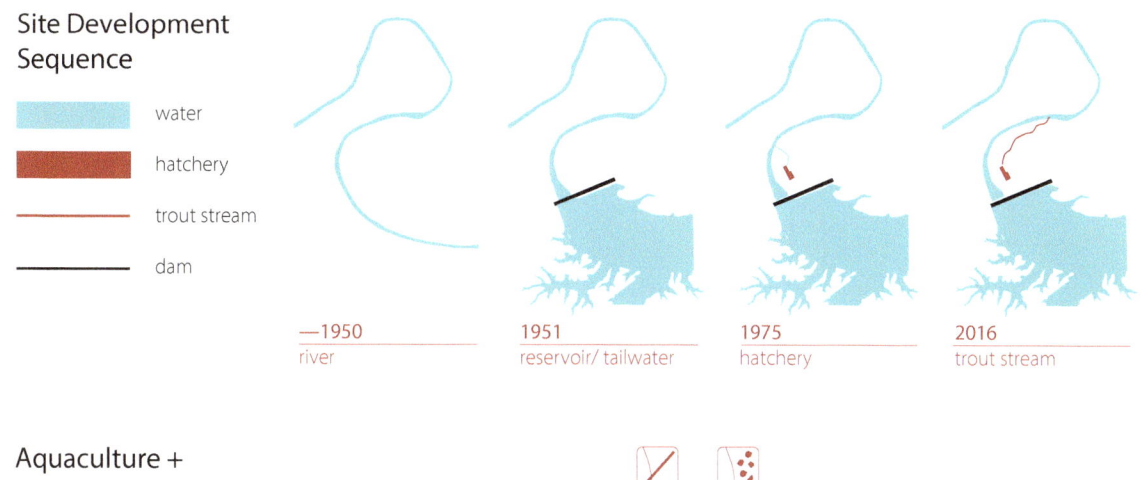

—1950	1951	1975	2016
river	reservoir/ tailwater	hatchery	trout stream

Aquaculture + Angling (2019)

- ▬ water
- ▬ buildings
- ─── roads
- ///// dam
- Type C stream · Rosgen classification
- ▤ toe wood
- ▨ riffle
- ▨ vane

Type DA stream

Type C stream

Tailwater
egg | alevin | fry | juvenile | adult | spawn

Trout Stream
egg | alevin | fry | juvenile | adult | spawn

Hatchery
egg | alevin | fry | juvenile | adult | spawn

07 *Landscape Systems and Strategies.* The tailwater, trout stream, and hatchery are elements of an aquaculture landscape that has been evolving for decades. Different trout life stages are enacted at each of these elements, but Hatchery Creek features Type A, C and DA classifications and constructed habitat designed to support the full range of trout life stages.

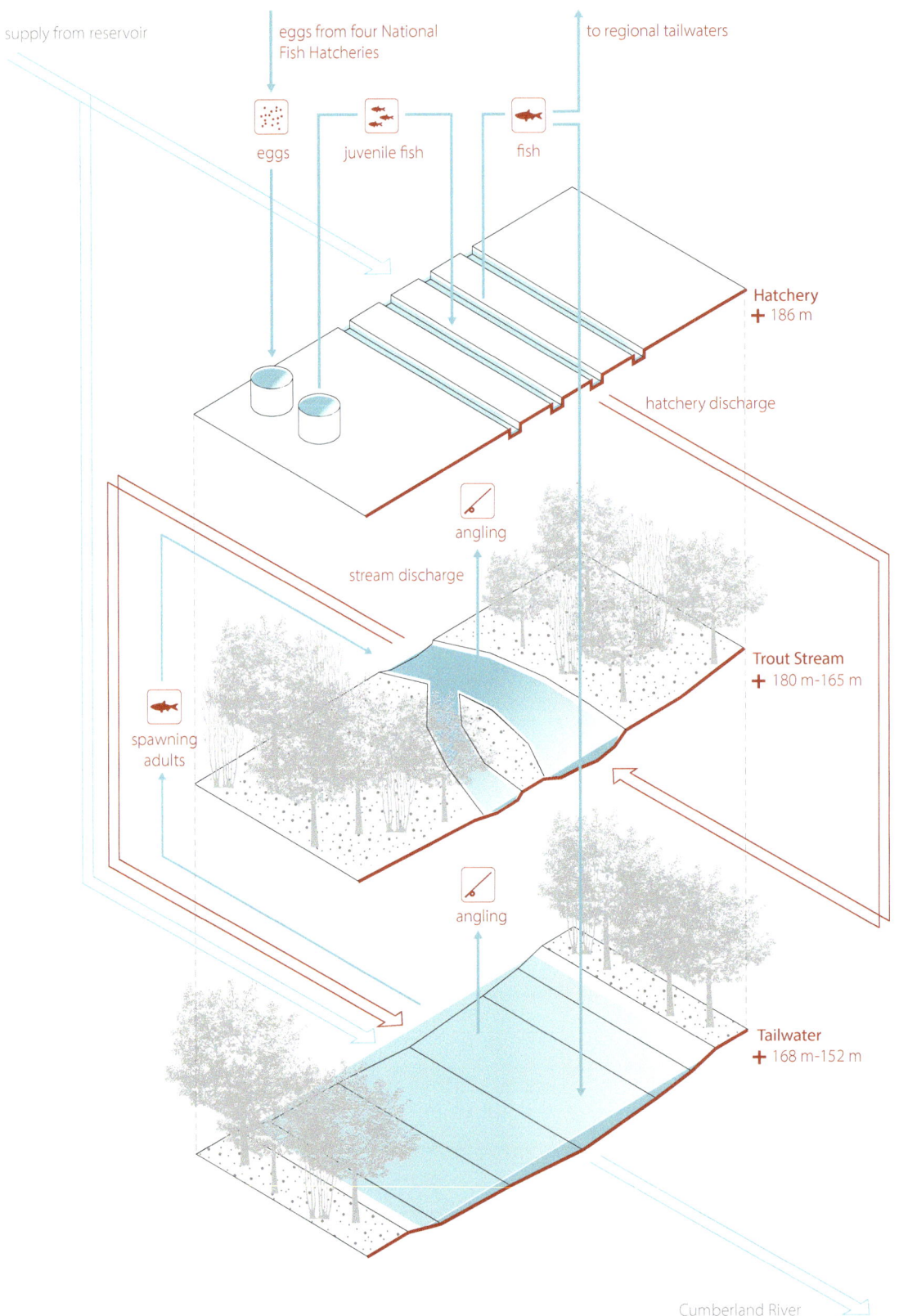

supply from reservoir

eggs from four National
Fish Hatcheries

to regional tailwaters

eggs

juvenile fish

fish

Hatchery
+ 186 m

hatchery discharge

angling

stream discharge

Trout Stream
+ 180 m-165 m

spawning
adults

angling

Tailwater
+ 168 m-152 m

Cumberland River

08 *Landscape Systems and Strategies.* Exchanges of water and fish occur between the hatchery, trout stream, and tailwater. Water discharged from the hatchery provides consistent flow of cold water to the trout stream, which ultimately flows into the Cumberland River. Some fish produced at the hatchery and stocked in the river travel to the trout stream to spawn.

09 Migration barrier at Hatchery Creek with bridge beyond.
10 View to the discharge of Hatchery Creek into the Cumberland River, nearly three kilometers downstream of the Wolf Creek Dam. The step ponds in the stream allow spawning trout to ascend the 10 m elevation gain to access the stream.

01 Aerial view of Ahuapaʻa ʻO Kahana and the Koʻolau mountains beyond. Huilua Fishpond, a loko kuapa (coastal fishpond) is seen in the foreground in Kahana Bay.

Ahupua'a of the Hawaiian Islands, United States

Landscape Type: ahupua'a

Landscape Area: 2,145 ha (*Ahupua'a 'O Kahana*, 19th C)
                      2,685 ha (Hawaiian fishponds, 19th C)

Aquaculture Yield: 336 kg/ha/yr (Hawaiian Islands, 19th C)

Aquaculture Type: extensive

Water Type: fresh/ brackish/ salt

On the pre-colonial Hawaiian Islands, between the thirteenth and nineteenth centuries, Hawaiians subdivided land into *ahupua'a*, a unit of land that typically encompassed a watershed extending from mountain ridges to a depth of two meters into the sea. Ahupua'a featured integrated aquaculture, agriculture, and silviculture that supported a Hawaiian family group and a local chief and included dispersed settlements and temples.

Aquaculture scholar Barry Costa-Pierce notes that the integrated farming system employed at the ahupua'a "spanned the normal salinity range of water and comprised a continuum from agriculture to aquaculture."[1] Five types of fishponds were constructed at ahupua'a, including freshwater, brackish, and seawater ponds. These fishponds were: *loko i'a kalo* (freshwater ponds for taro and fish cultivation), *loko wai* (freshwater fishponds), *loko pu'uone* (brackish coastal fishponds), *loko kuapa* (shoreline ponds enclosed by rock seawalls), and *loko 'umeiki* (shoreline fishtraps). Prior to 1900, the island of O'ahu featured the highest number of fishponds of any Hawaiian island. At least 360 ponds, yielding an estimated 900,000 kg of fish per year, operated on this island.[2]

The aquaculture systems of the ahupua'a featured several inventive practices and technologies. Upland fishponds were supplied with freshwater and interconnected via an extensive system of channels with variable-flow rates called *'auwai*. Another innovation was planting taro on small mounds within loko i'a kalo, which created channels for fish that fed on insects and taro leaves. The coastal loko kuapa featured perhaps the most remarkable practices. A channel constructed in the thick rock wall of the fishpond connected the pond to the sea and enabled both fish stocking and harvesting. A stationary *makaha* (wooden grate) in the channel allowed small fish to enter the pond, but it trapped larger fish growing in the pond.[3] During spawning seasons the makaha was lifted and hundreds of fish attempting to leave through the channel were easily netted.

Ahupua'a 'O Kahana is an intact 2,100 ha ahupua'a on the island of O'ahu. Today it is managed as a state park. It is estimated that up to 1,000 indigenous people lived in this ahupua'a at the time of Western contact in 1776. Today, approximately thirty people live on the ahupua'a and interpret the site for visitors. During a period of intense cultivation of this ahupua'a (1200-1800 CE), a total of 120 loko i'a kalo were constructed for fish and taro production.[4] At the coast, in the Bay of Kahana, a 3 ha loko kuapa has been restored. Restoration of loko kuapa walls across the Hawaiian Islands are often communal, public efforts that involve conservation groups, hundreds of volunteers, and employ traditional construction methods. These events connect contemporary island inhabitants to the traditional aquaculture landscapes of the Hawaiian Islands.

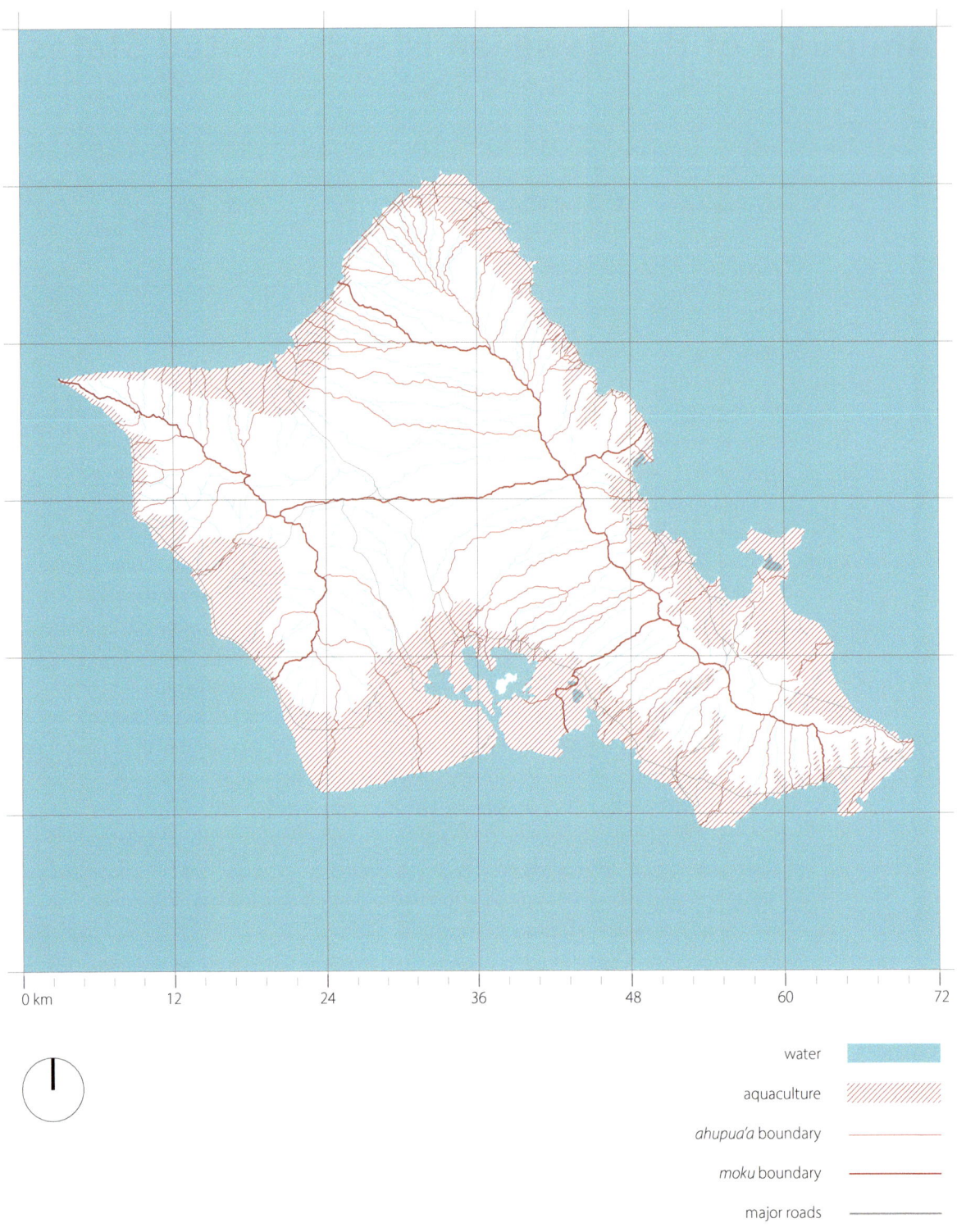

water

aquaculture

ahupua'a boundary

moku boundary

major roads

0 km   12   24   36   48   60   72

02 *Regional Diagram.* A diagram of O'ahu Island, circa 1885, depicts native Hawaiian land divisions called *moku* and *ahupua'a.* O'ahu is estimated to have had 175 fishponds prior to 1900, totaling over 1,300 ha. The aquaculture zone depicted here extends up to an elevation of 60 m. Freshwater fishponds were generally not built above this elevation.

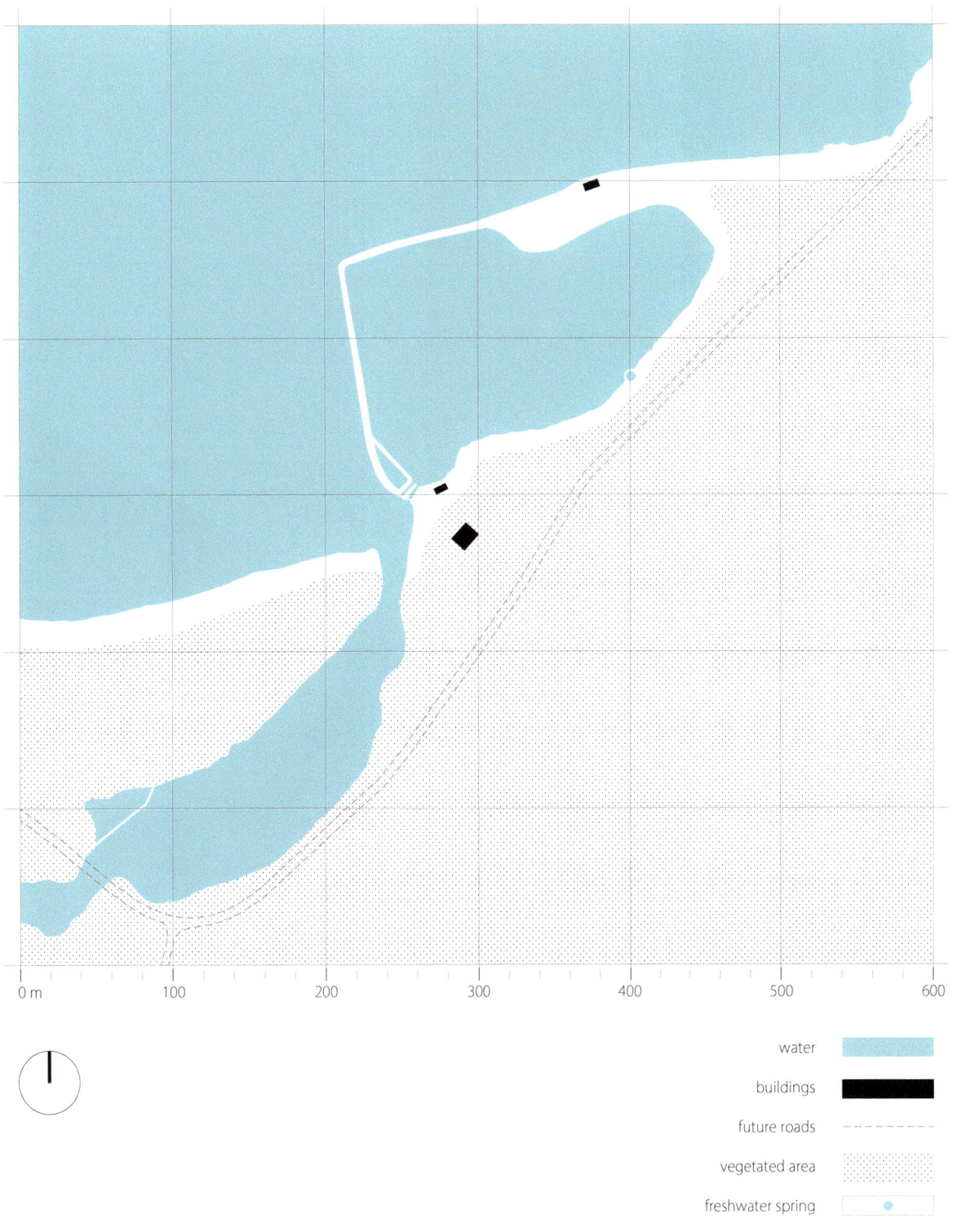

water	
buildings	
future roads	- - - - - - -
vegetated area	
freshwater spring	

03 *Site Diagram.* Huilua Fishpond, a loko kuapa used for passive fish capture and fish rearing at the bay of Ahupua'a 'O Kahana, depicted circa 1885. The water at the loko kuapa is brackish as it is fed by both tidal bay waters and a freshwater spring. A loko wai (freshwater pond) is located along the banks of Kahana Stream.

40

0m   0m

Cultivated Fauna

Striped Mullet *(Mugil cephalus)*
Silver perch *(Kuhlia sandwicensis)*
Milkfish *(Chanos chanos)*
Bonefish *(Albula vulpes)*
Ladyfish *(Elops hawaiensis)*
Pacific Threadfin *(Polydactylus sexfilis)*
moray eels *(Gymnothorax spp.)*
Parrot Fish *(Scaridae)*
unicornfish *(Naso spp.)*

Hawaiian Gobies *(Eleotris sandwicensis)*
Tahitian prawn *(Macrobrachium spp.)*

Cultivated Flora

taro *(Colocasia esculenta)*
yam *(Dioscorea spp.)*
banana *(Musa spp.)*
breadfruit *(Artocarpus altilis)*
coconut *(Cocos nucifera)*

hale kia'i (pond keeper's house)

pond for raising fry

kuapa (seawall)

'auwai (canal)

makaha (wooden grate)

04   *Transect.* At the loko kuapa at Ahupua'a 'O Kahana, tidal water enters the pond through two stationary makahas (wooden grates) that allow water and small fish to enter, but prevent larger fish in the pond from escaping. The seawall of the loko kuapa, constructed using coral and lava rock, doubles near the makahas to create a pond used for raising *pua* (fry).

05 Volunteers participate in the restoration and reconstruction of the seawall of a loko kuapa on the island of Molokai.
06 Aerial view of the Huilua Fishpond at Ahupua'a 'O Kahana. The channel of this loko kuapa through which fish once entered the pond is currently buried under earth near to where the seawall connects to land.

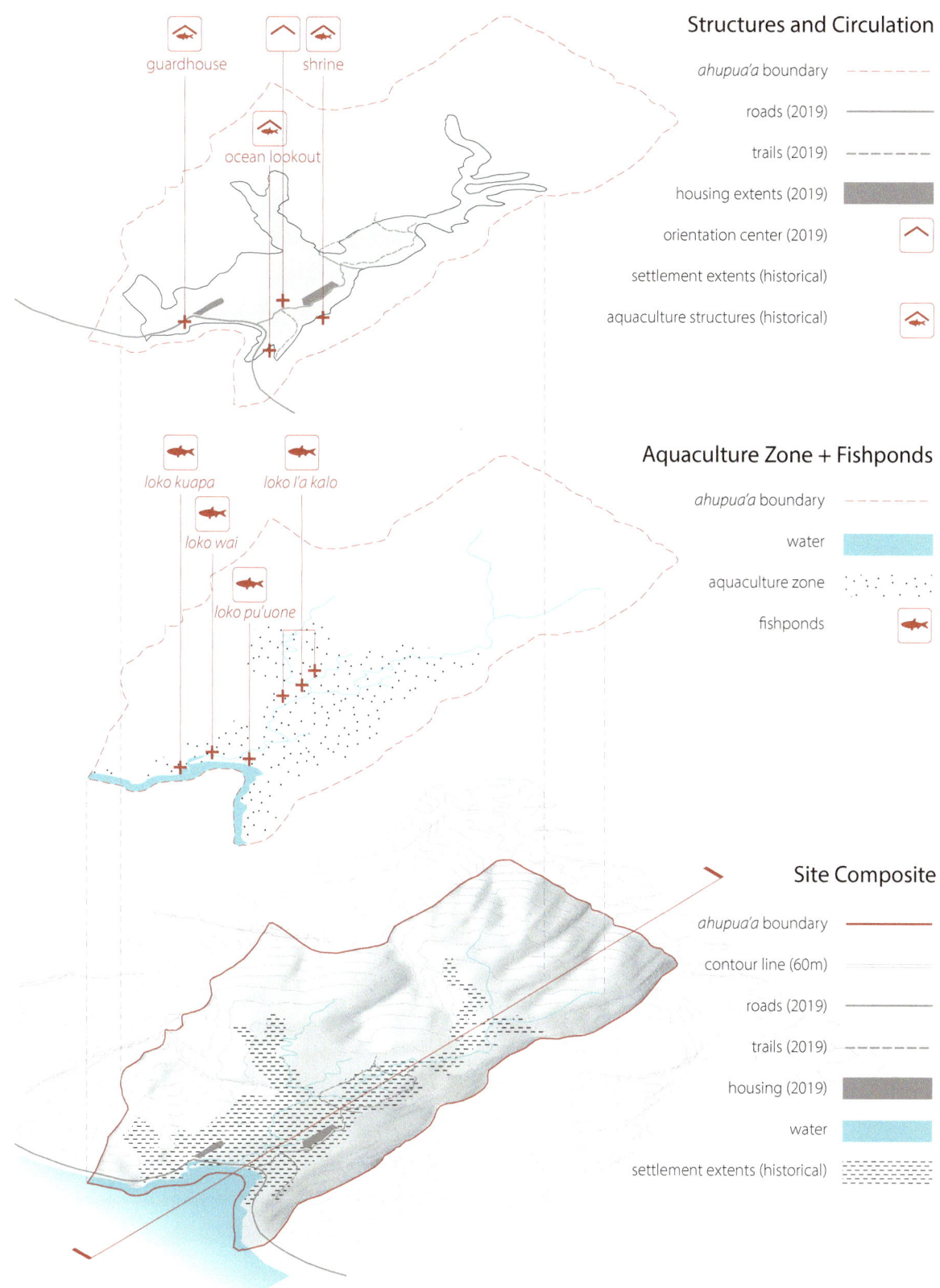

Structures and Circulation

guardhouse
shrine
ocean lookout

ahupua'a boundary	
roads (2019)	
trails (2019)	
housing extents (2019)	
orientation center (2019)	
settlement extents (historical)	
aquaculture structures (historical)	

Aquaculture Zone + Fishponds

loko kuapa
loko l'a kalo
loko wai
loko pu'uone

ahupua'a boundary	
water	
aquaculture zone	
fishponds	

Site Composite

ahupua'a boundary	
contour line (60m)	
roads (2019)	
trails (2019)	
housing (2019)	
water	
settlement extents (historical)	

07 *Landscape Systems and Strategies.* Aquaculture, agriculture, and housing settlements were intermixed in the valley of the Ahupua'a 'O Kahana. Historically there were a variety of sacred and utilitarian buildings associated with various sites for aquaculture in the ahupua'a, including ocean fish lookouts, guardhouses, and shrines.

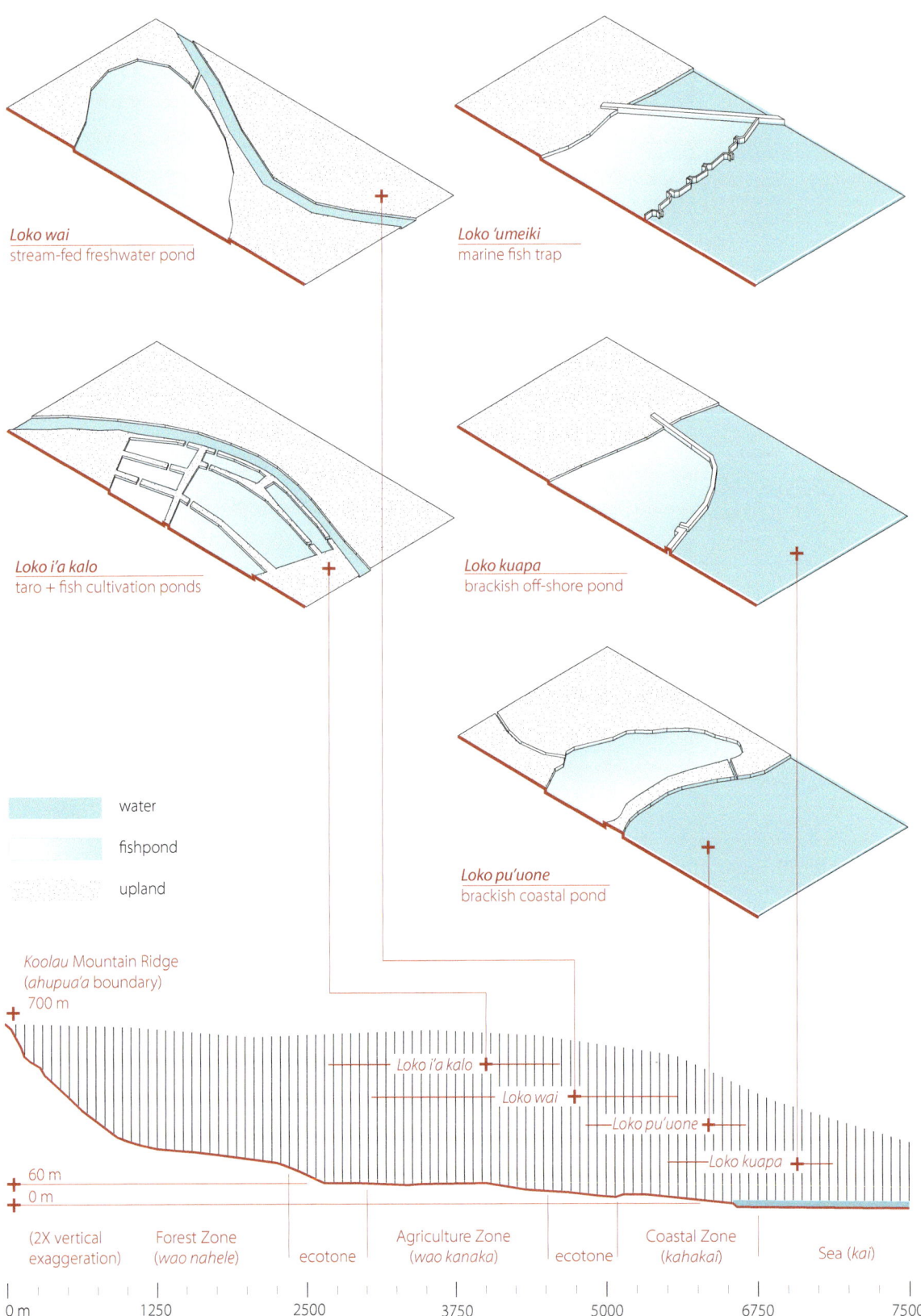

Loko wai
stream-fed freshwater pond

Loko ʻumeiki
marine fish trap

Loko iʻa kalo
taro + fish cultivation ponds

Loko kuapa
brackish off-shore pond

water

fishpond

upland

Loko puʻuone
brackish coastal pond

Koolau Mountain Ridge
(*ahupuaʻa* boundary)
700 m

Loko iʻa kalo

Loko wai

Loko puʻuone

Loko kuapa

60 m

0 m

(2X verticlal
exaggeration)

Forest Zone
(*wao nahele*)

ecotone

Agriculture Zone
(*wao kanaka*)

ecotone

Coastal Zone
(*kahakai*)

Sea (*kai*)

0 m   1250   2500   3750   5000   6750   7500

08 *Landscape Systems and Strategies.* Ahupuaʻa ʻO Kahana is a watershed that is subdivided into ecological zones based on vegetation and elevation. Four types of ponds for aquaculture and agriculture were constructed below an elevation of 60 m. The ponds allowed for the cultivation of fish in freshwater, brackish water, and sea water.

09 View to loko iʻa kalo, terraced freshwater ponds constructed for the integrated cultivation of taro and fish, at Limahuli Garden and Preserve on Kauaʻi Island. In ponds like these, fish would passively feed on taro leaf and insects and were also fed seaweed, cut grass, and shellfish.

PART THREE

Depicting Aquaculture Landscapes

01 SCAPE, *Oyster-tecture*, physical model, 2009-2010. The model depicts a proposed living reef landscape in New York Harbor inhabited by humans, oysters, mussels, shorebirds, and eelgrass. The lack of water plane implies continuity between benthic habitat, emergent islands, and recreation zones (see Part One).

Representations of Aquaculture Landscapes

Depictions of aquaculture have accompanied fish farming practices since their earliest iterations. But evocative, in situ representations of fish by early humans precede the advent of farming. A vivid and detailed sculpture of a spawning Atlantic Salmon, carved to scale in a cave now known as *Abri du Poisson* in France, dates to around 25,000 BCE.[1] Another example, found deep in *Cueva de la Pileta* in Spain, is the elegant, 1.5 m long "Great Black Fish" cave painting, which dates to ~18,000 BCE.[2] These depictions of fish are intrinsic to the limestone landscapes in which they were created. As John Berger imagines, the Paleolithic cave artist, "with red pigment on his finger, could persuade [animals] to come to the rock's surface, to brush against it and stain it with their smells."[3]

Contemporaneous representations of historical aquaculture landscapes and practices are plentiful. Bas relief on the Tomb of Thebaine from ~2,000 BCE in the Nile Delta, Egypt, depicts tilapia culture in an artificial pond.[4] Clay and stone models of rice fields with figurines of carp, frogs, turtles, and farmers, dating to the Han Dynasty (25-220 CE), have been unearthed in China.[5] Glass vases from ancient Rome from ~300 CE illustrate *ostrearia*, Latin for "oyster culture grounds," at which hanging-ropes were used for oyster cultivation.[6] Bird's-eye views of post-medieval English country houses, such as the 1712 engraving titled, *Dyrham the Seat of William Blathwayt Esq.*, by Johannes Kip, illustrated medieval fish ponds that were adapted into formal gardens and landscape parks.[7] In Japan, in the mid-nineteenth century, fishermen created evocative images of the fish they caught by coating the animals in ink and then pressing paper onto them. This method is called *gyotaku* in Japanese.[8] The birth of the modern science of aquaculture in France in the mid-nineteenth century was accompanied by the production and dissemination of treatises that featured detailed engravings of a range of aquaculture technologies and processes.[9] Journals from this era, such as *L'Illustration Journal Universelle* in France, brought aquaculture into the public imagination with engravings of fantastic hatcheries, aquaria, and aquatic exhibitions. Today, contemporary landscape practitioners, academics, and artists contribute to this history of representation through their own depictions of aquaculture landscapes.

The following pages contain a curated collection of twenty-five drawings, maps, paintings, collages, models, and sculptures. Many of these visualizations document existing landscapes, while others are speculative. They were created using analog, digital, or hybrid methods. Some of these visualizations were commissioned by municipalities and institutions whereas others were initiated organically through independent research. Three-dimensional models and sculptures are joined by two-dimensional site plans and diagrams. Some drawings illustrate an ichthyocentric perspective, and others represent the perspectival vantage of humans, birds, and satellites. The subjects of these works range in scale from the length of an eel to the breadth of a continent.

James Corner writes that landscape drawing is "fundamentally an eidetic and generative activity, one where the drawing acts as a producing agent or ideational catalyst."[10] The creative visualizations of real and imagined aquaculture landscapes that are collected in this section reveal drawing and modeling as ongoing, fertile processes through which we continue to discover our relationships to farmed fish.

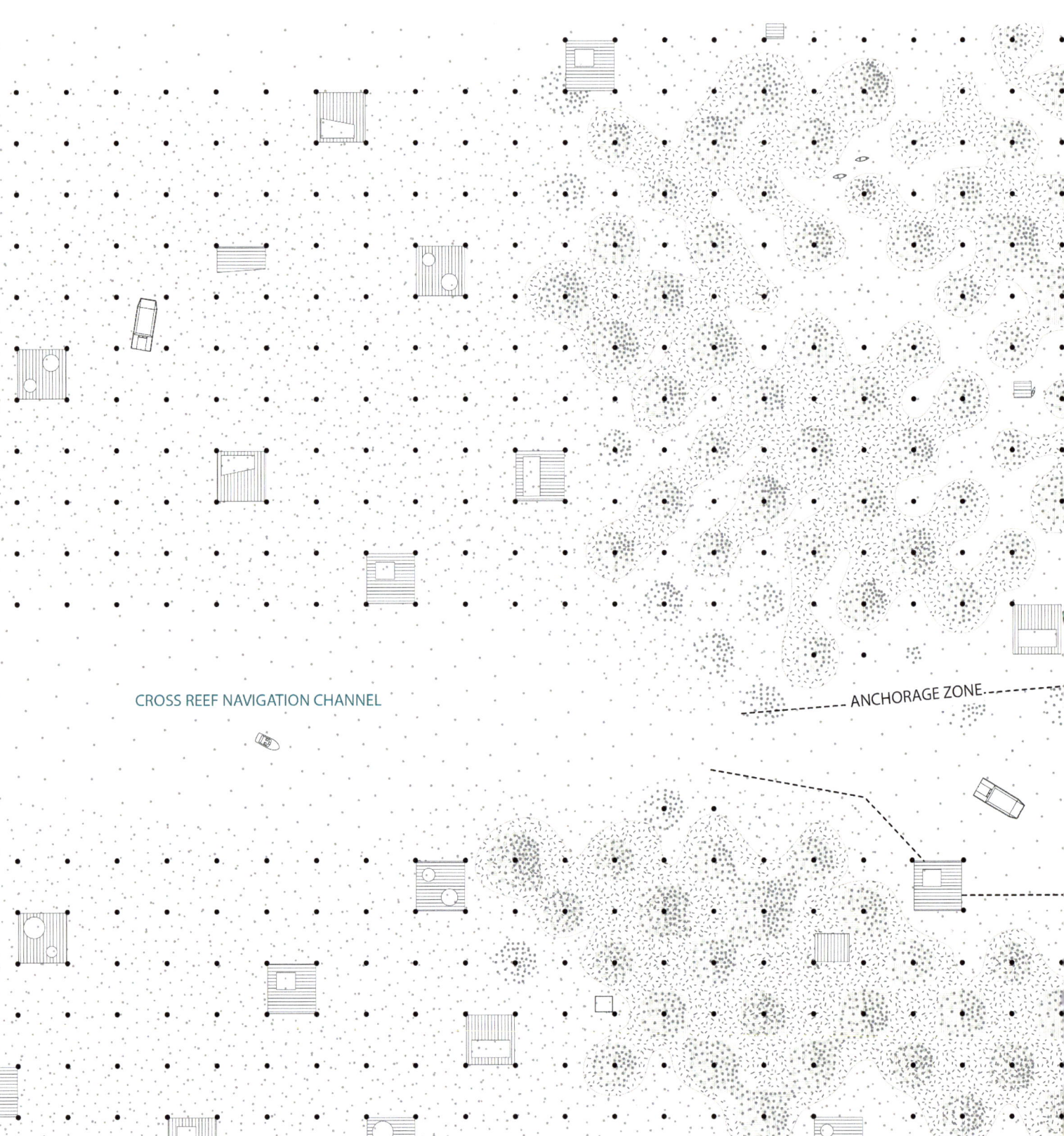

CROSS REEF NAVIGATION CHANNEL

ANCHORAGE ZONE

02 SCAPE, *Oyster-tecture*, digital drawing, 2009-2010. Plan of proposed Palisades Reef in New York Harbor, a constructed reef characterized by a field of pylons, nets, docks, diving platforms, and intermittent islands. Navigation channels create boat access through this archipelago (see Part One).

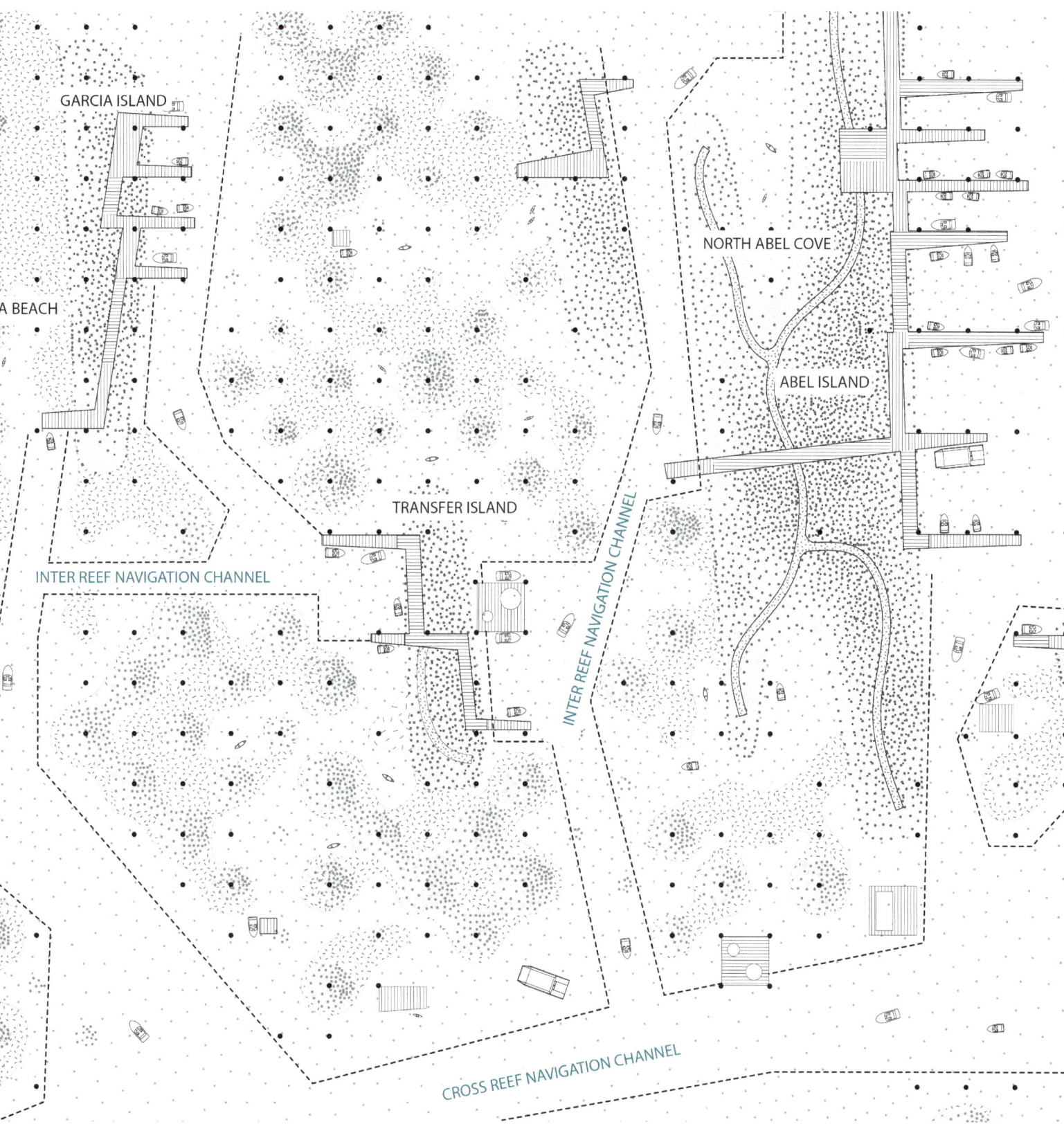

GARCIA ISLAND

A BEACH

NORTH ABEL COVE

ABEL ISLAND

TRANSFER ISLAND

INTER REEF NAVIGATION CHANNEL

INTER REEF NAVIGATION CHANNEL

CROSS REEF NAVIGATION CHANNEL

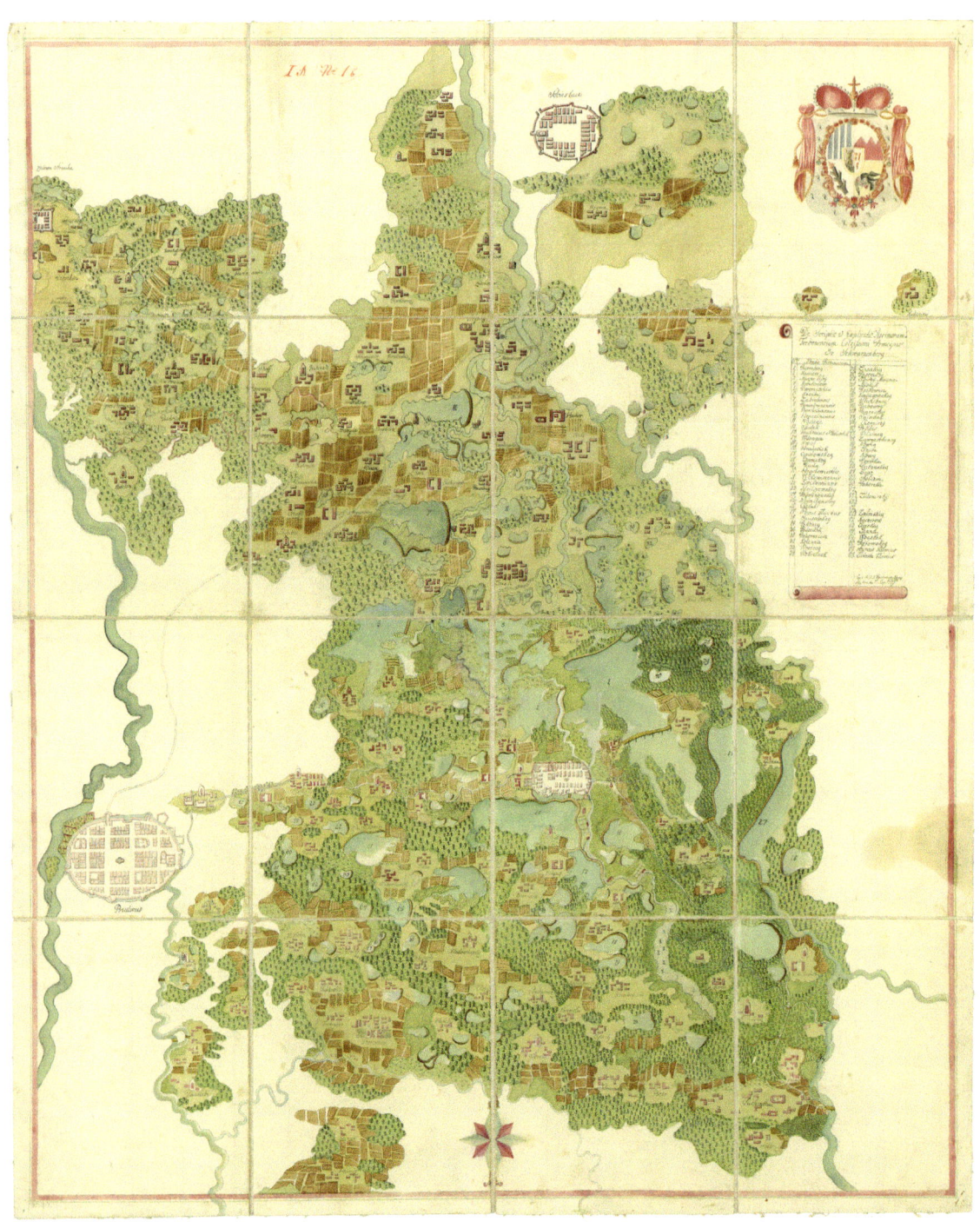

03 W. J. J. Pachmann, *Mapa rybniční soustavy třeboňského panství* (*Map of the pond system in the domain Třeboň*), 1779. Plan of Třeboň Basin in the Czech Republic reveals a landscape characterized by hundreds of constructed fishponds (see Case Study 01).

04 *Mapa rybniční soustavy třeboňského panství (detail).* Ponds constructed for fish farming are numbered. The brown, sinuous edges at the fishponds depict earthen dams. The town of Třeboň is at the center of the image.

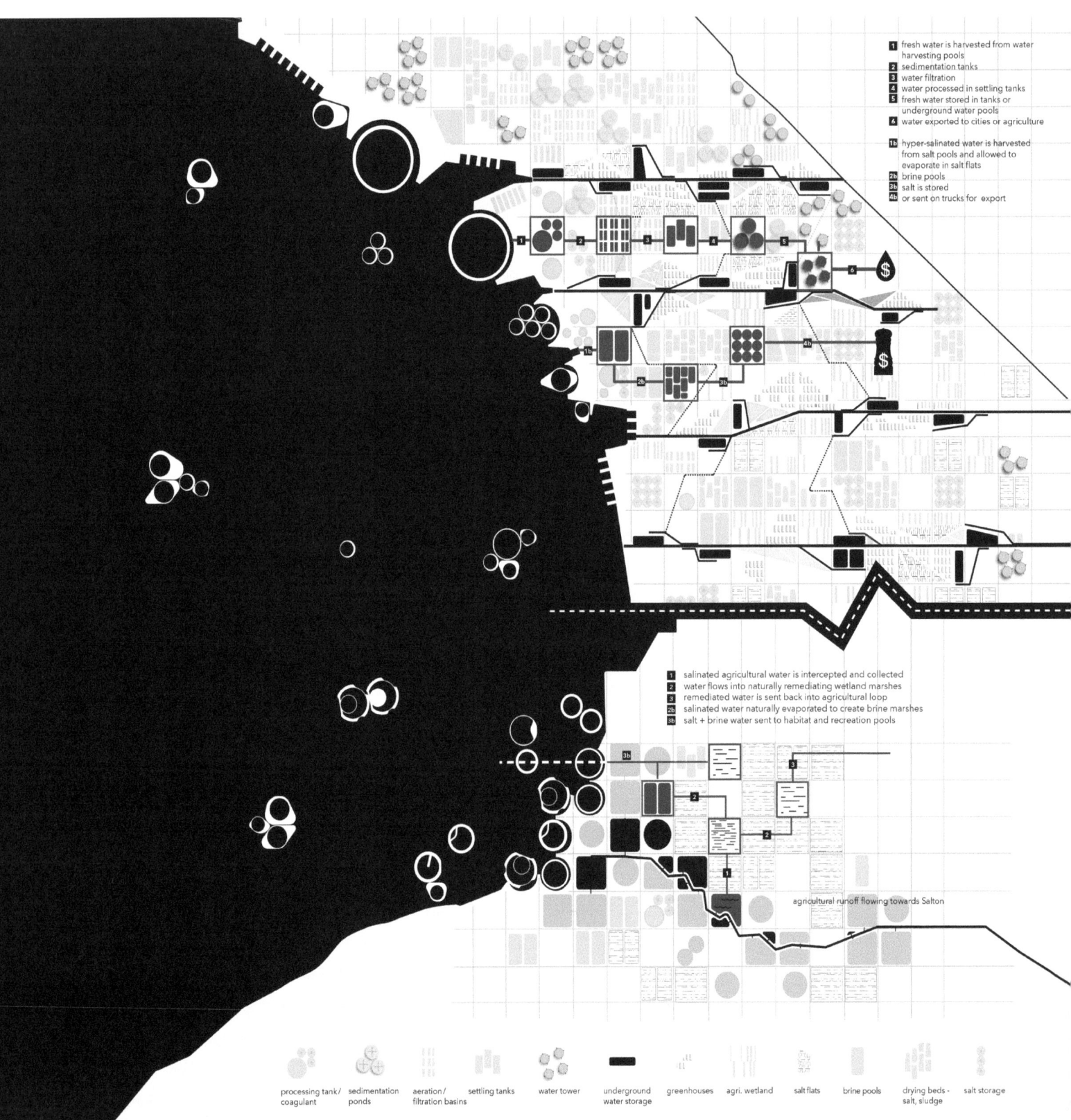

1 fresh water is harvested from water harvesting pools
2 sedimentation tanks
3 water filtration
4 water processed in settling tanks
5 fresh water stored in tanks or underground water pools
6 water exported to cities or agriculture

1b hyper-salinated water is harvested from salt pools and allowed to evaporate in salt flats
2b brine pools
3b salt is stored
4b or sent on trucks for export

1 salinated agricultural water is intercepted and collected
2 water flows into naturally remediating wetland marshes
3 remediated water is sent back into agricultural loop
2b salinated water naturally evaporated to create brine marshes
3b salt + brine water sent to habitat and recreation pools

agricultural runoff flowing towards Salton

processing tank/ coagulant | sedimentation ponds | aeration/ filtration basins | settling tanks | water tower | underground water storage | greenhouses | agri. wetland | salt flats | brine pools | drying beds - salt, sludge | salt storage

05 Lateral Office, *Water Economies/Water Ecologies*, digital drawing, 2010. Site plan of southeastern shore of the Salton Sea in California, USA. A mosaic of basins that support remedial and productive processes is proposed for the shore of this evaporating lake. Floating pools aggregate across the Salton Sea and support aquaculture and other programs.

Aquaculture/Kelp: Cluster
1. Salt harvester and algae nursery feed brine shrimp
2. Brine shrimp feed fish culture
3. Fish culture waste is cleaned by kelp nursery
4. Kelp is brought onto pool platforms to dry in sun
5. Fish and kelp are brought to shore for harvest

Habitat: Cluster
1. Salton Sea is a major bird migratory area
2. Habitat mass is created by clustering of pools
3. Brine shrimp and algae pools provide food for birds

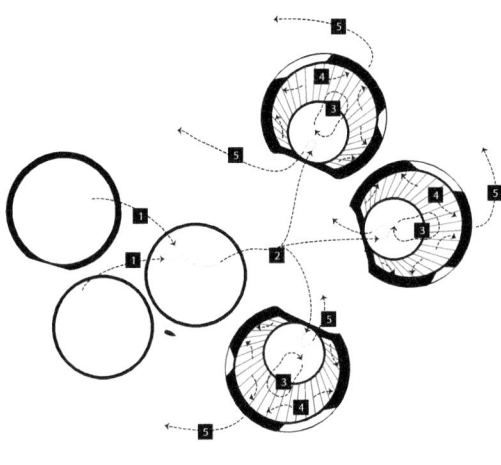

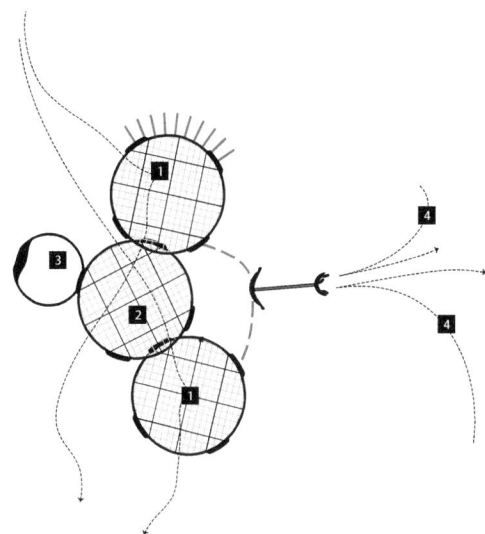

Recreation: Cluster
1. Boats pull up to recreation pool
2. Freshwater pools deliver recreation pool
3. Saltwater feeds therapeutic pools

Salt/Water: Cluster
1. Water pool takes in saline sea water
2. Desalinated water delivered to shore for treatment
3. Brine pools deliver saltwater for salt harvesting

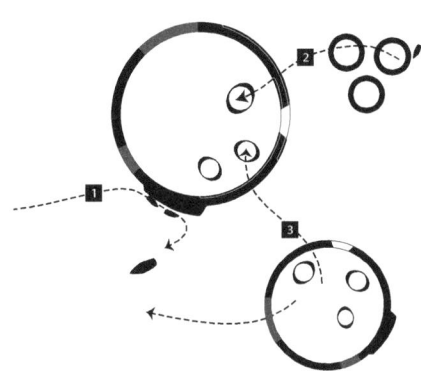

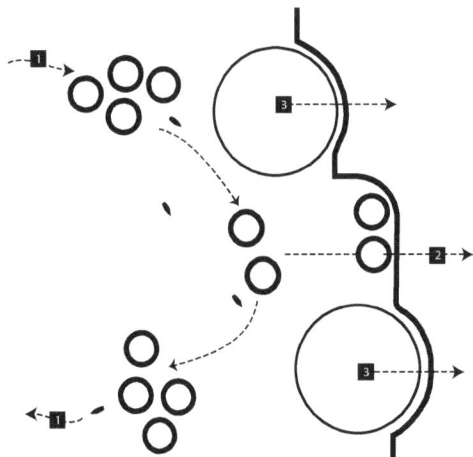

06 *Water Economies/Water Ecologies*. Proposed floating pools of varying salinity support a range of productive, ecological, and recreational functions. Pools can aggregate into clusters that enable sequences and expand scales of production. For instance, algae and salt-harvester pools facilitate brine-shrimp production, which in turn services fish and kelp polyculture.

07 Forbes Lipschitz and Justine Holzman, detail from *The Secret Lives of Catfish: Waterfowl on an Active Catfish Pond*, digital composite, 2016. This digital drawing imagines multispecies cohabitation at an active catfish farm in the Mississippi River Delta. Fish farms in this historically flood-prone region hold water and provide habitat for migrating waterfowl.

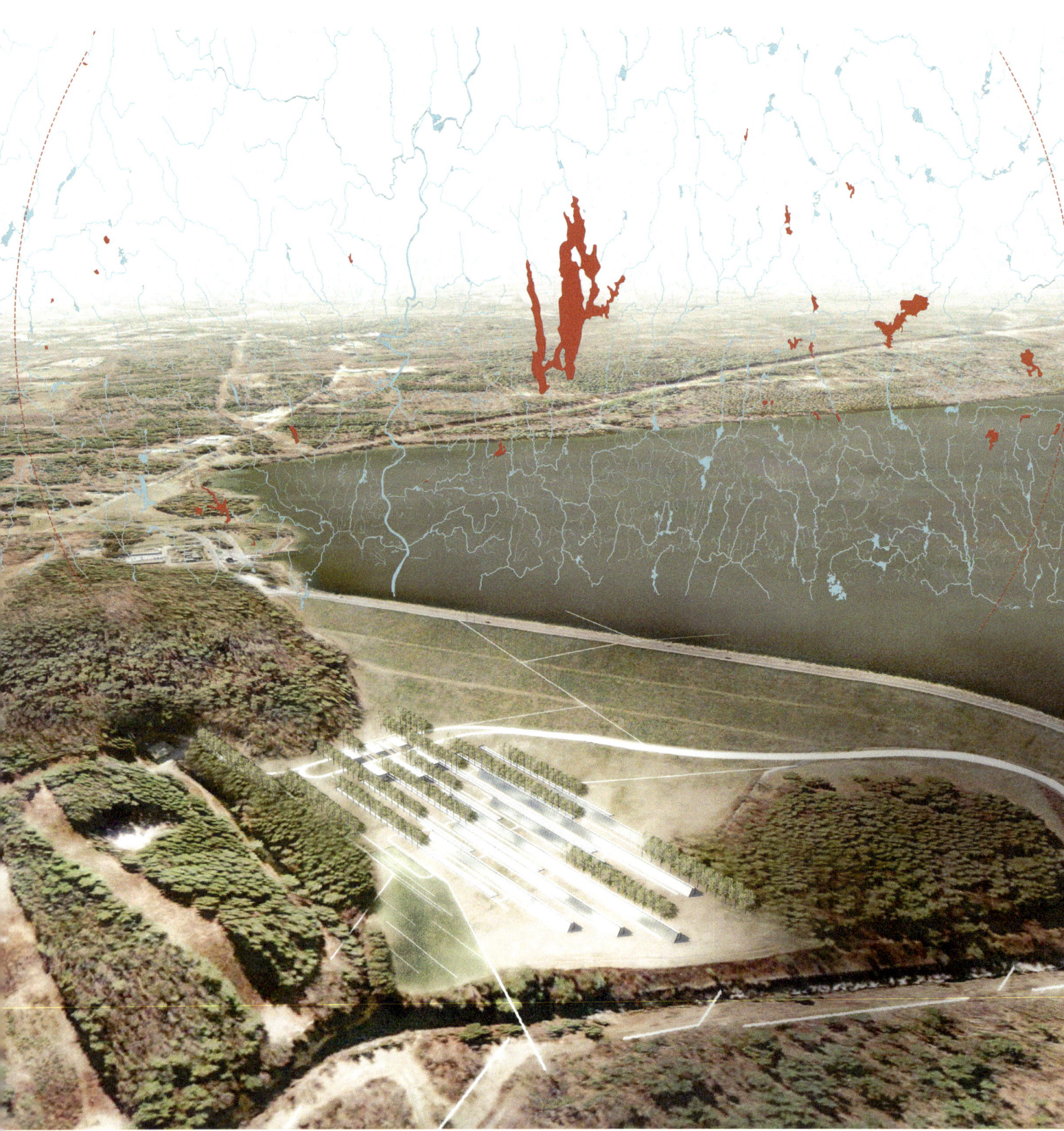

08 Michael Ezban, *Quabbin Fishery*, digital composite, 2014. Aerial oblique that combines satellite imagery and digital modeling of a proposed aquaponic landscape at the base of Winsor Dam at Quabbin Reservoir in Massachusetts, USA. GIS imagery of regional reservoirs and tailwaters stocked with fish by state hatcheries overlays this collage (see Part One).

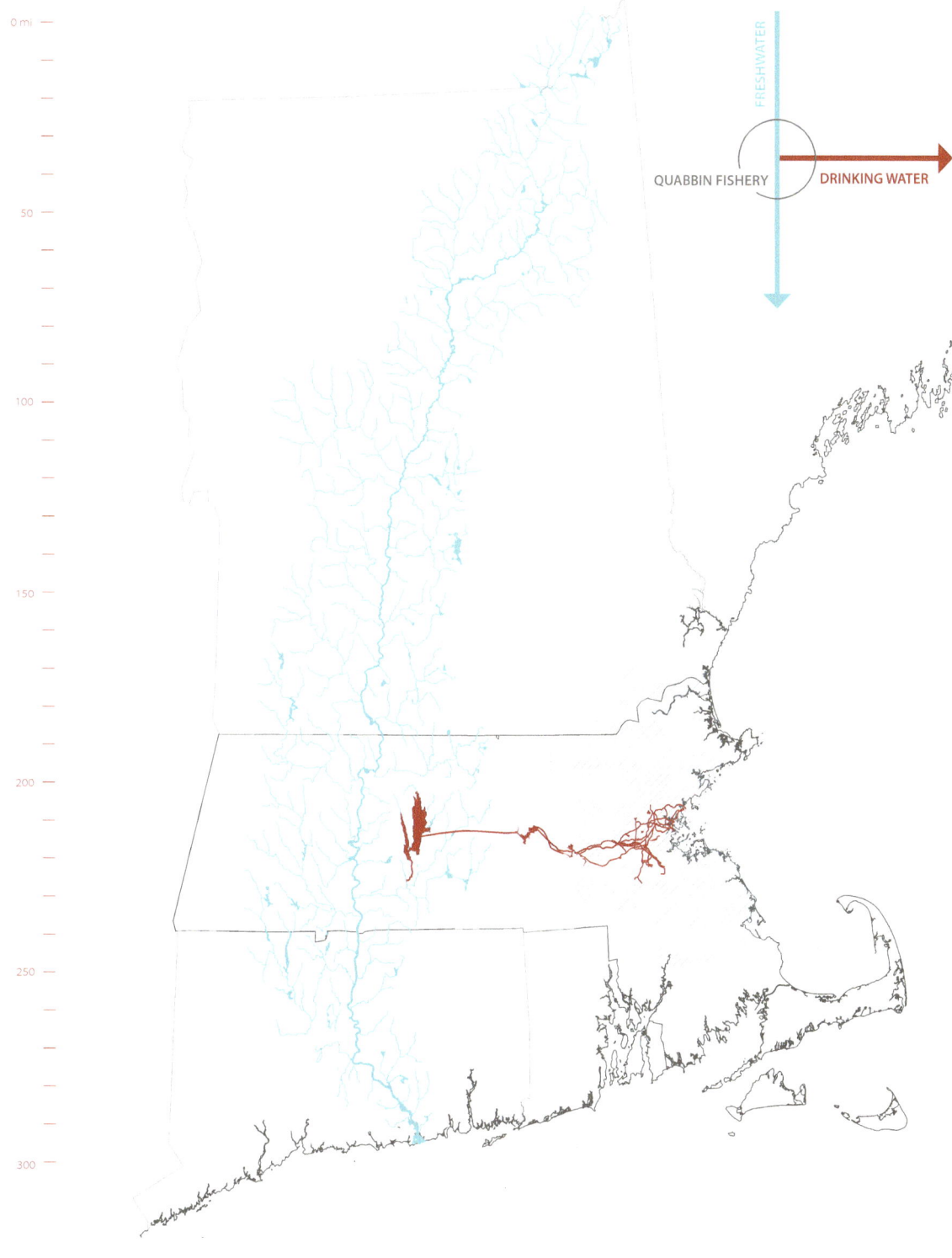

09 *Quabbin Fishery*. Aqueducts convey drinking water from Quabbin Reservoir to Boston. The Connecticut River and its tributaries carry freshwater to Long Island Sound. Quabbin Fishery is sited at the only point of intersection between these two aquatic systems—the Swift River, the tailwater of the Quabbin Reservoir, which discharges into the Connecticut River.

10 Artist unknown, detail from *Plan Théorique de la Lagune de Comacchio*, 1861. A diagram of the Valli di Comacchio in Italy, a landscape constructed to capture catadromous eels. The diagram depicts the elaborate branching system of (a) dikes, and (b) channels. Salt water is rendered dark blue; fresh and brackish water is light blue (see Case Study 04).

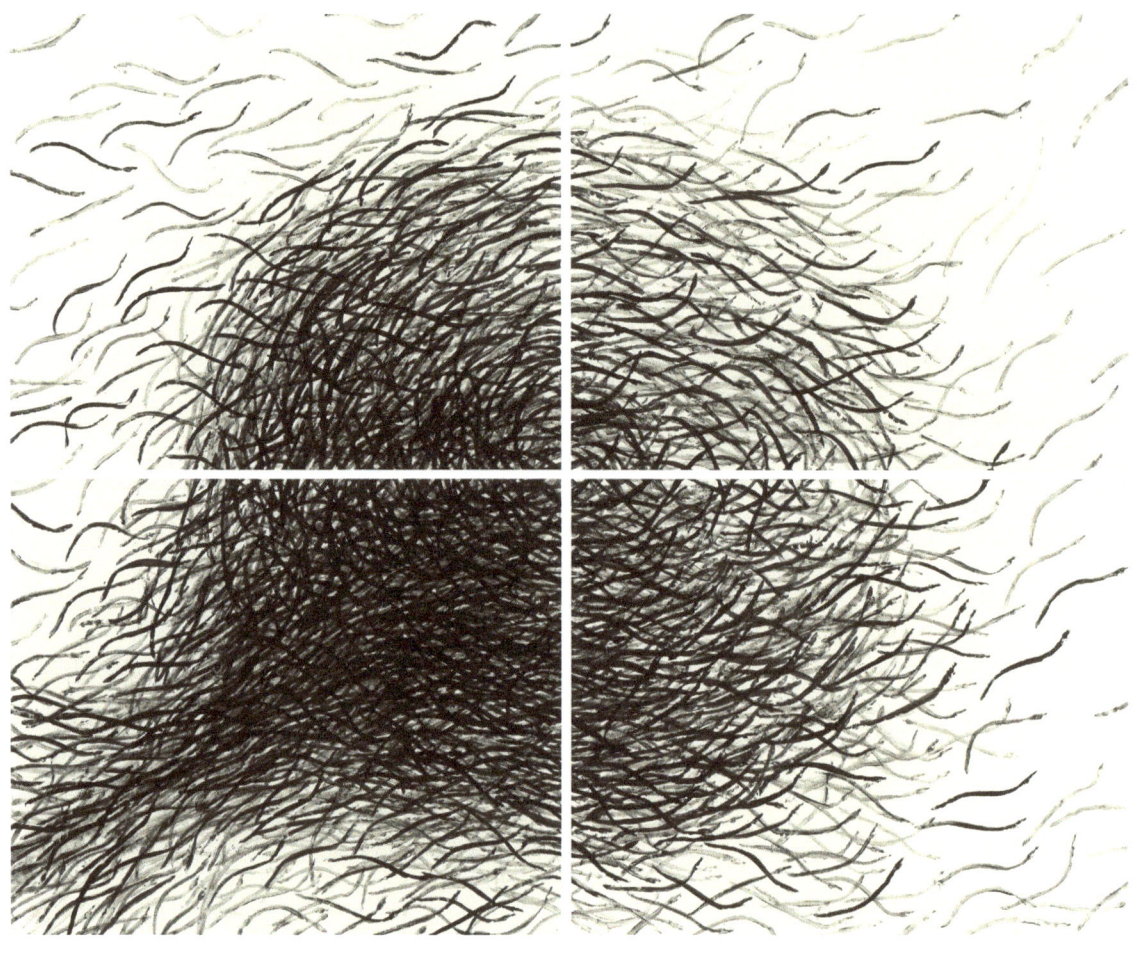

11 James Prosek, *Abstract Nature*, ink on paper, 2012. This image is a vision by artist James Prosek of eels spawning in the Sargasso Sea, an enigmatic zone bounded only by ocean currents. After leaving this hatching ground some young eels are caught along coastlines and raised on farms. Adult eels are also harvested as they travel back to the Sargasso Sea to spawn.

12 *Abstract Nature (detail)*. The image is created using the *gyotaku* technique—dead eels were painted with ink, and then their bodies were pressed onto paper. The process developed in Japan in the nineteenth century as a means for fishermen to document their catches. This detail is at a quarter of actual scale.

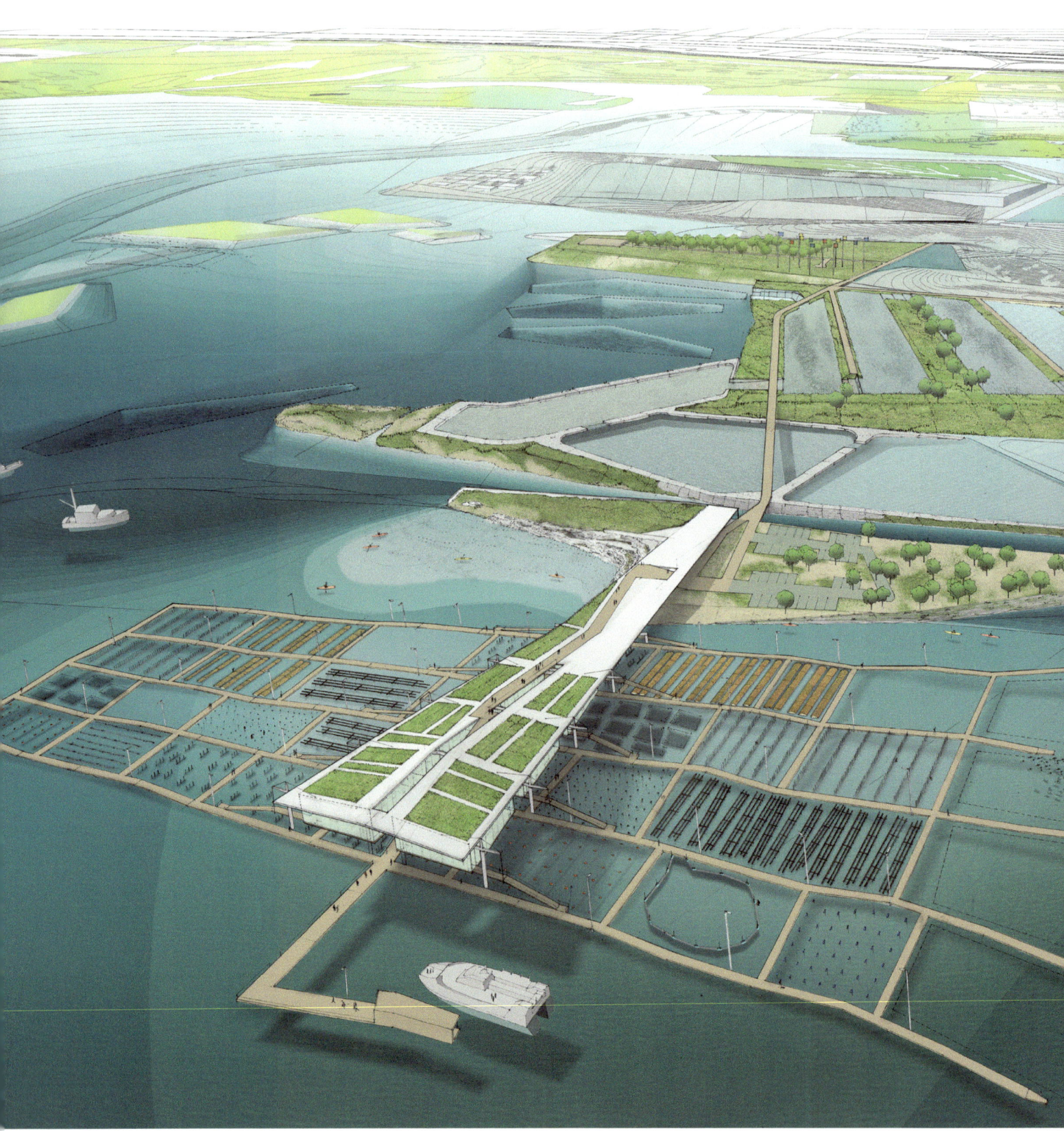

13 LTL Architects, detail from *Water Proving Grounds*, digital and analog drawing, 2010. Proposal for productive, intertidal landscapes in response to sea-level rise in New York Harbor. View to a multifunctional land pier that features an aquaculture-based program including a research and development facility and hatchery, as well as recreational pools and parks beyond.

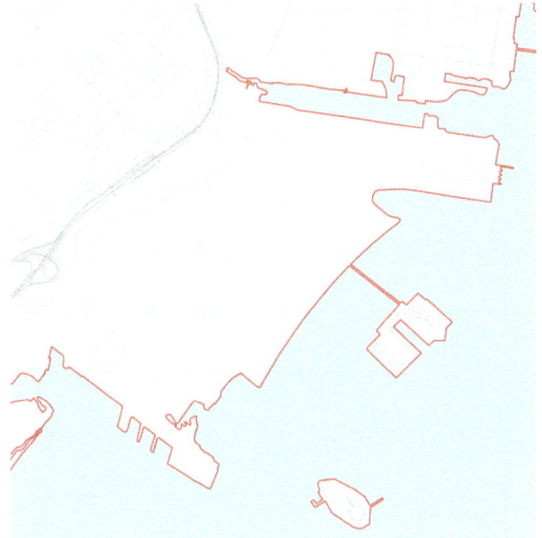

Existing: five miles of coast

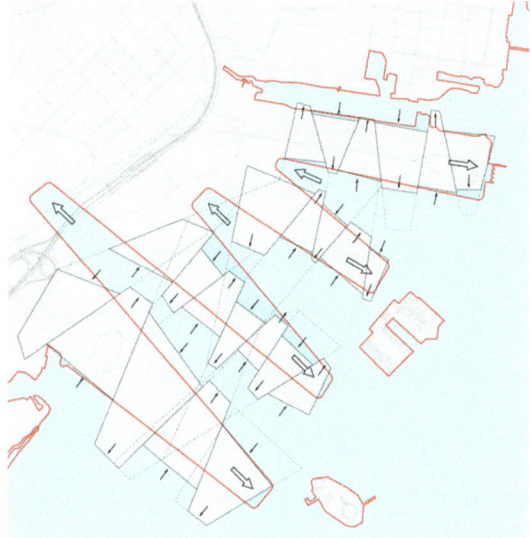

Proposed: piers and crossgrain

Circulation and anchors

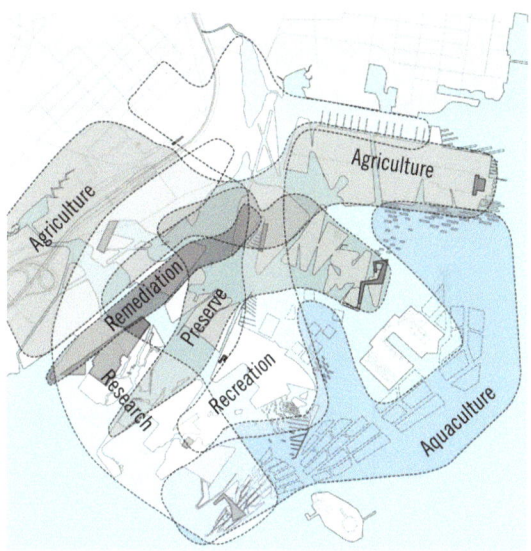

Program areas

14 *Water Proving Grounds.* Plan diagrams describe the proposal for four land-based piers that increase the length of the shoreline at the site by a factor of ten. Aquaculture programs overlap agriculture, recreation, and research program zones, and extend public experiences into intertidal and benthic environments between and beyond the piers.

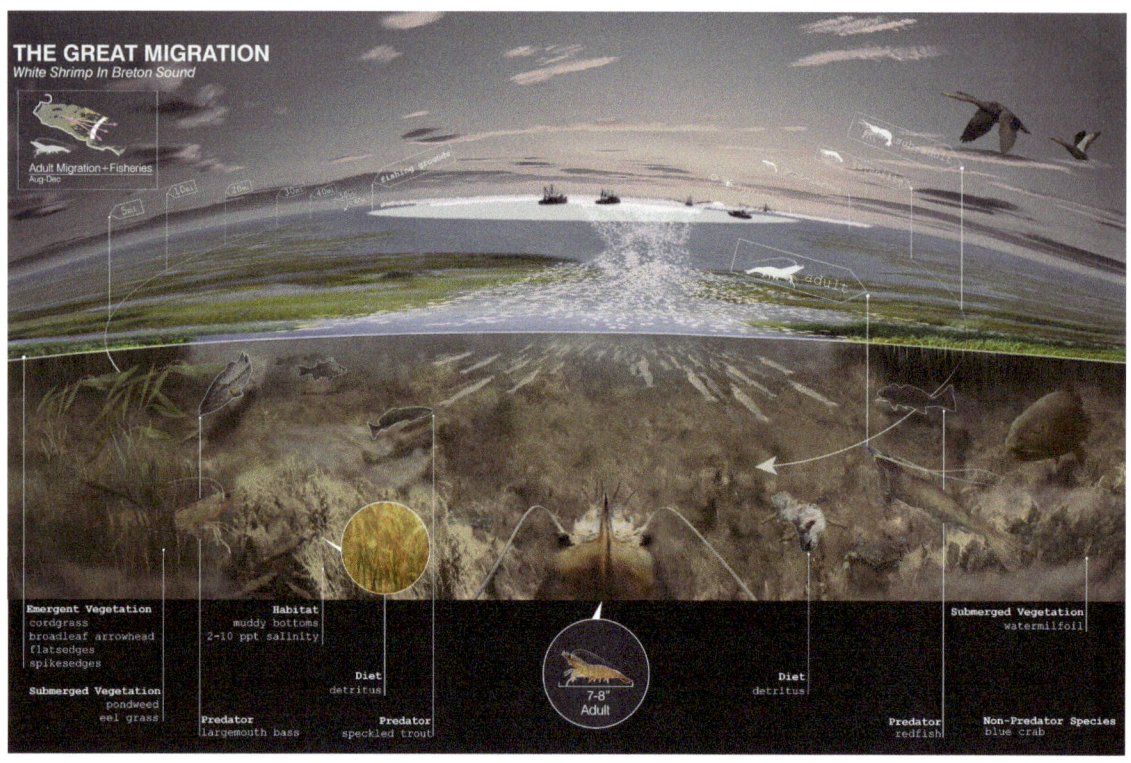

The Great Migration
White Shrimp In Breton Sound

15 Louisiana State University Coastal Sustainability Studio, *Shrimp Eye View*, digital composite, 2016. This collage depicts a White Shrimp *(Litopenaeous setiferus)* in its freshwater habitat in Breton Sound, Louisiana. The drawing describes metrics and lifecycle stages associated with White Shrimp spawning migration.

16 *Shrimp Eye View (detail).* The drawing illustrates an ichthyocentric perspective and plunges the viewer into the water at Breton Sound. The view is a panorama of White Shrimp habitat, food sources, and predators—both fish and humans. A commercial fishing fleet is positioned to intercept shrimp during their 100 km migration to spawn in the Gulf of Mexico.

17 E. Simon, detail from *Plan Général; Etablissement de Pisciculture de Huningue,* lithograph,1866. This site plan of the piscifactoire at 1:2,000 scale (shown at full size) identifies the location, function, and size of fifteen distinct fishpond types, fifteen water channel types, and six water basin types. Arrows indicate direction of water flow (see Case Study 06).

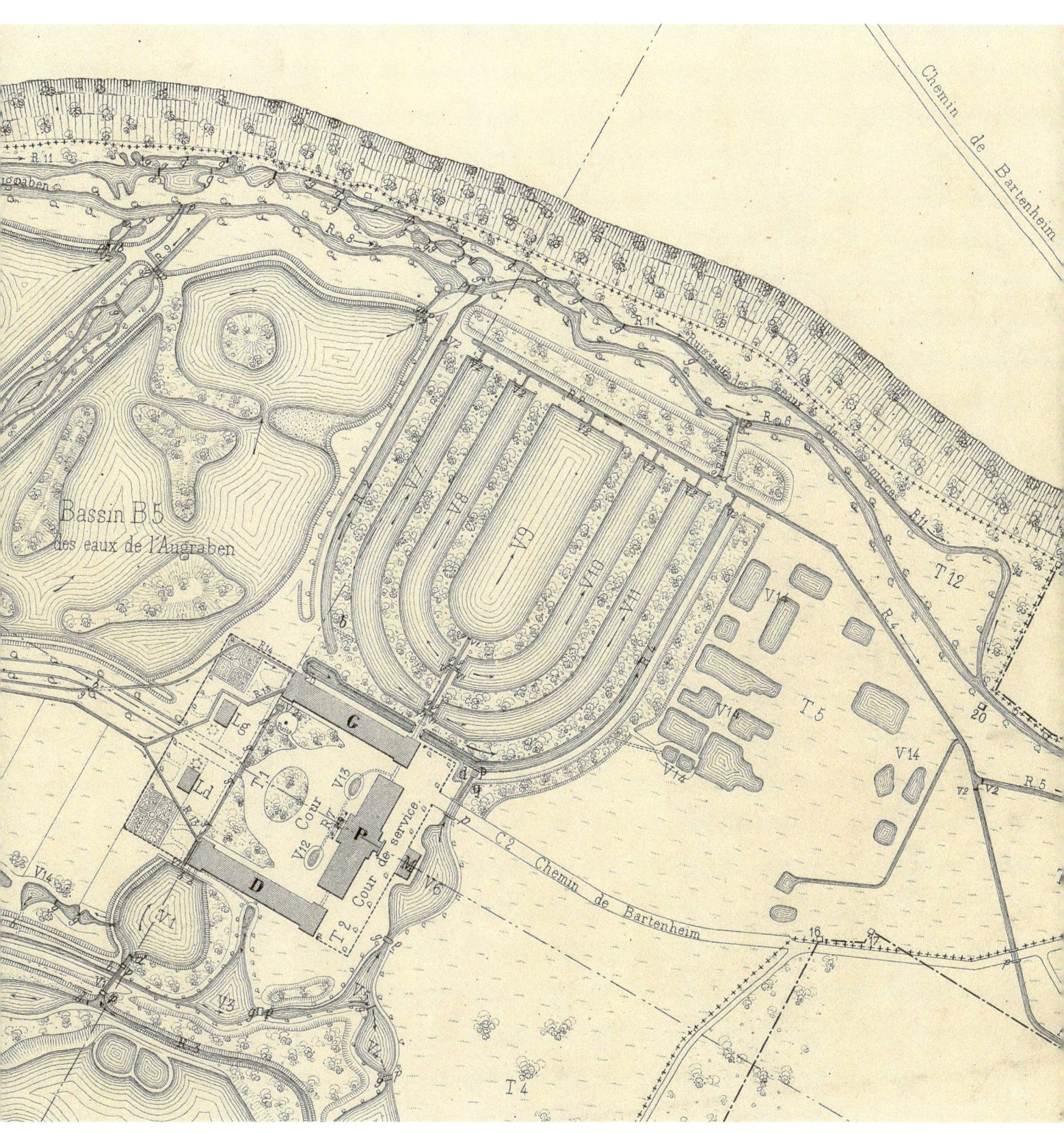

18 Mason Jackson, *Experimenting Ponds at the Huningue Fish Ponds*, etching, 1861. An etching of the piscifactoire after a photograph by Adolphe Braun, an influential French photographer of the Alsace region. This etching appeared in the London Illustrated News, a popular journal with a distribution of approximately 300,000 per week (see Case Study 06).

19 Michael Ezban, *Aqueous Ecologies*, digital rendering, 2013. Aerial oblique of ichthyological urbanism at Willets Point, a derelict peninsula in Queens, New York. Parametric modeling is used to develop iterations of a biodiverse urban framework of dendritic channels that support urban development and enable a multi-trophic aquaculture of fish, mussel, and kelp.

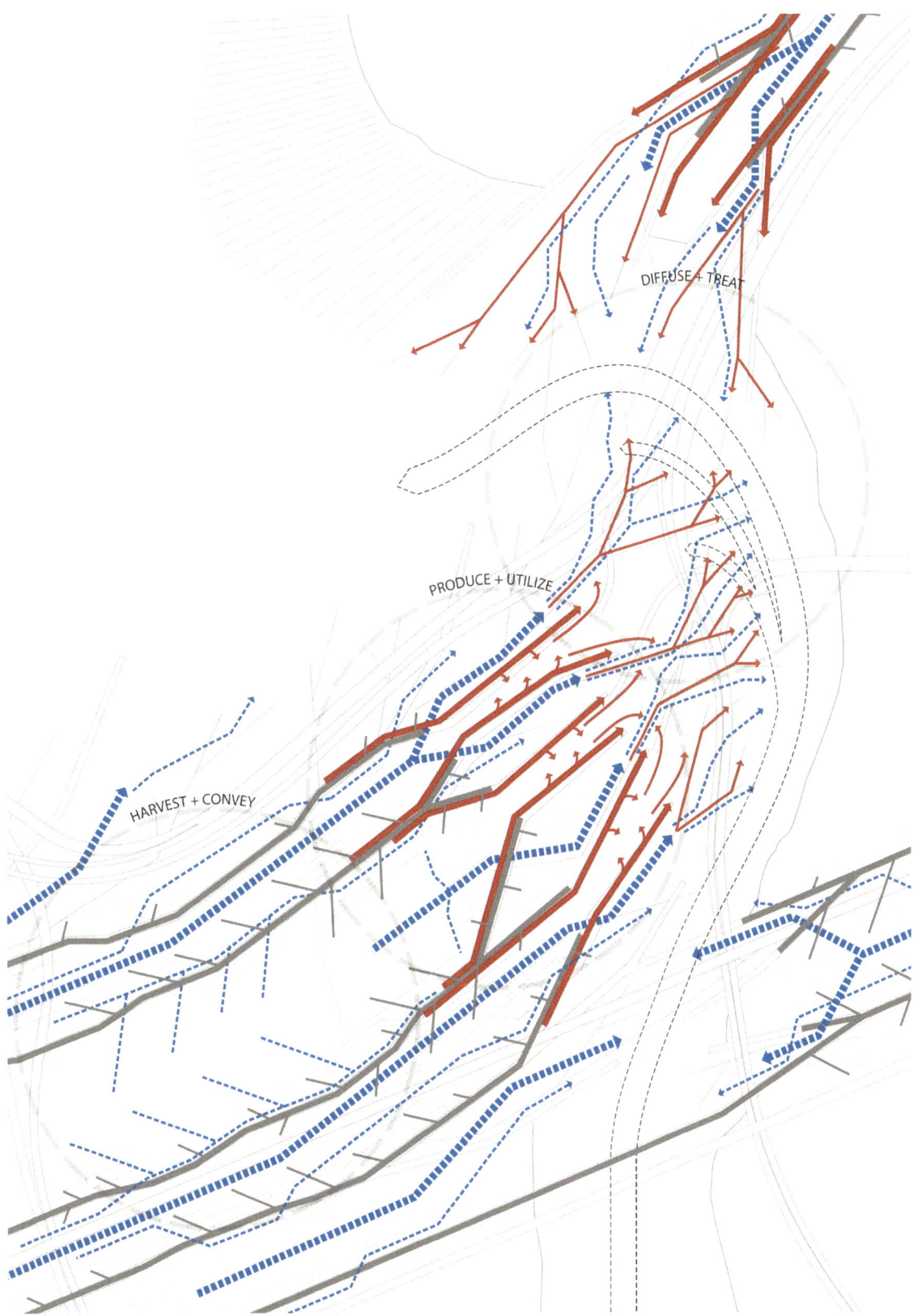

DIFFUSE + TREAT

PRODUCE + UTILIZE

HARVEST + CONVEY

20 *Aqueous Ecologies*. Greywater harvested from buildings is treated and directed to flow through fish rivulets, mussel beds, and kelp troughs. As local fish populations become self-sustaining, urban hatcheries transition into urban incubators, and kelp farms adapt into wetland waterfront. In ichthyological urbanism, fish have agency in guiding urban development.

21 SCAPE, detail from *Living Breakwaters*, digital composite, 2012. Breakwaters are conceptualized as living systems and habitat for a range of aquatic and avian species. This early concept image depicts habitat pockets called "reef streets" that are lined with special concrete units intended to increase biological recruitment (see Part One).

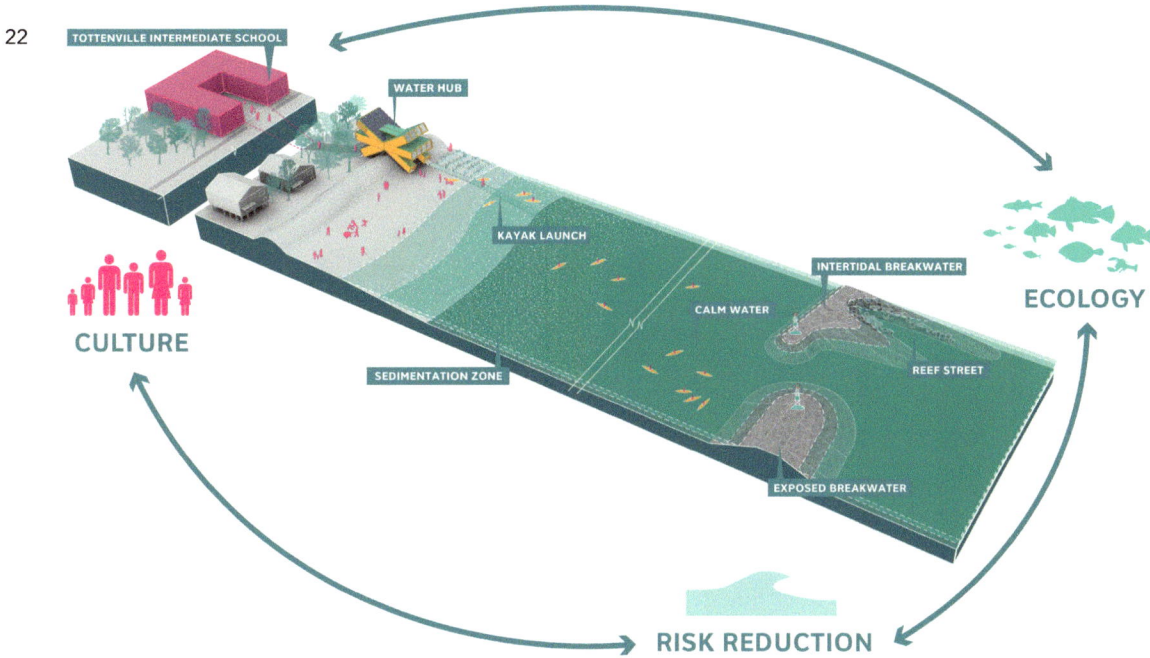

22

TOTTENVILLE INTERMEDIATE SCHOOL

WATER HUB

KAYAK LAUNCH

SEDIMENTATION ZONE

CALM WATER

INTERTIDAL BREAKWATER

REEF STREET

EXPOSED BREAKWATER

CULTURE

ECOLOGY

RISK REDUCTION

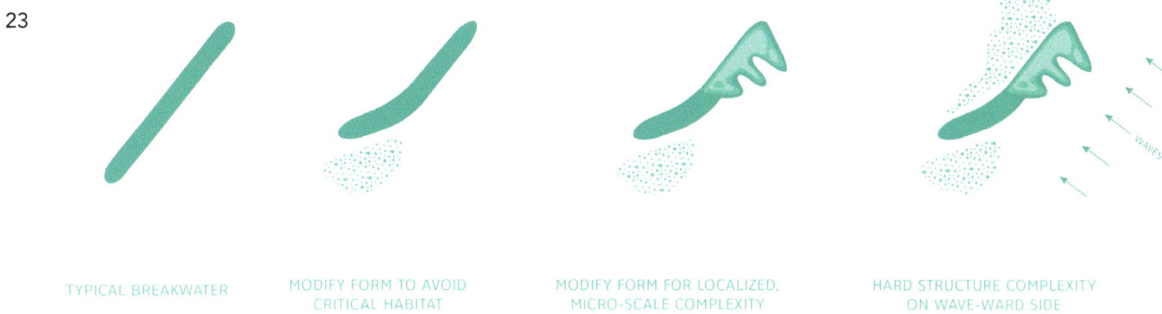

23

TYPICAL BREAKWATER

MODIFY FORM TO AVOID
CRITICAL HABITAT

MODIFY FORM FOR LOCALIZED,
MICRO-SCALE COMPLEXITY

HARD STRUCTURE COMPLEXITY
ON WAVE-WARD SIDE

WAVES

22 *Living Breakwaters*. Transect depicts a multipronged strategy for resiliency intended to dissipate destructive wave energy, revive marine and nearshore ecologies, and foster a culture of citizen stewardship, education, and recreation.
23 *Living Breakwaters*. Diagram of adaptations to a conventional, linear breakwater to create and protect habitat zones.

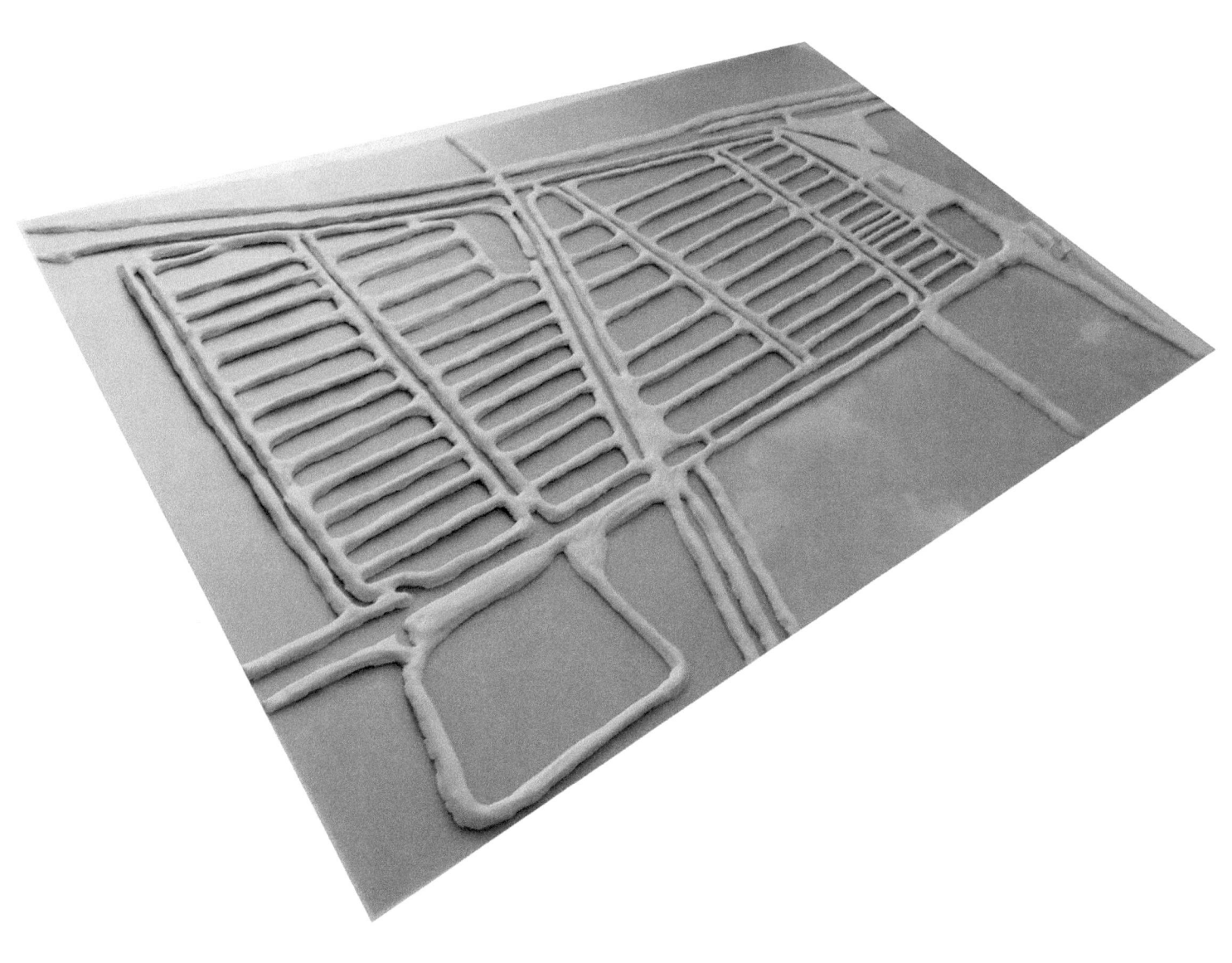

24 Estudi Martí Franch (EMF), *La Tancada Lagoon Park*, plasticine models, 2009-2010. This model depicts existing conditions at a former fish farm and salt works in the Ebro River delta in Spain. Salt had been produced here since the twelfth century. At the end of the twentieth century the salt pans were excavated into linear ponds that enabled a decade of fish farming.

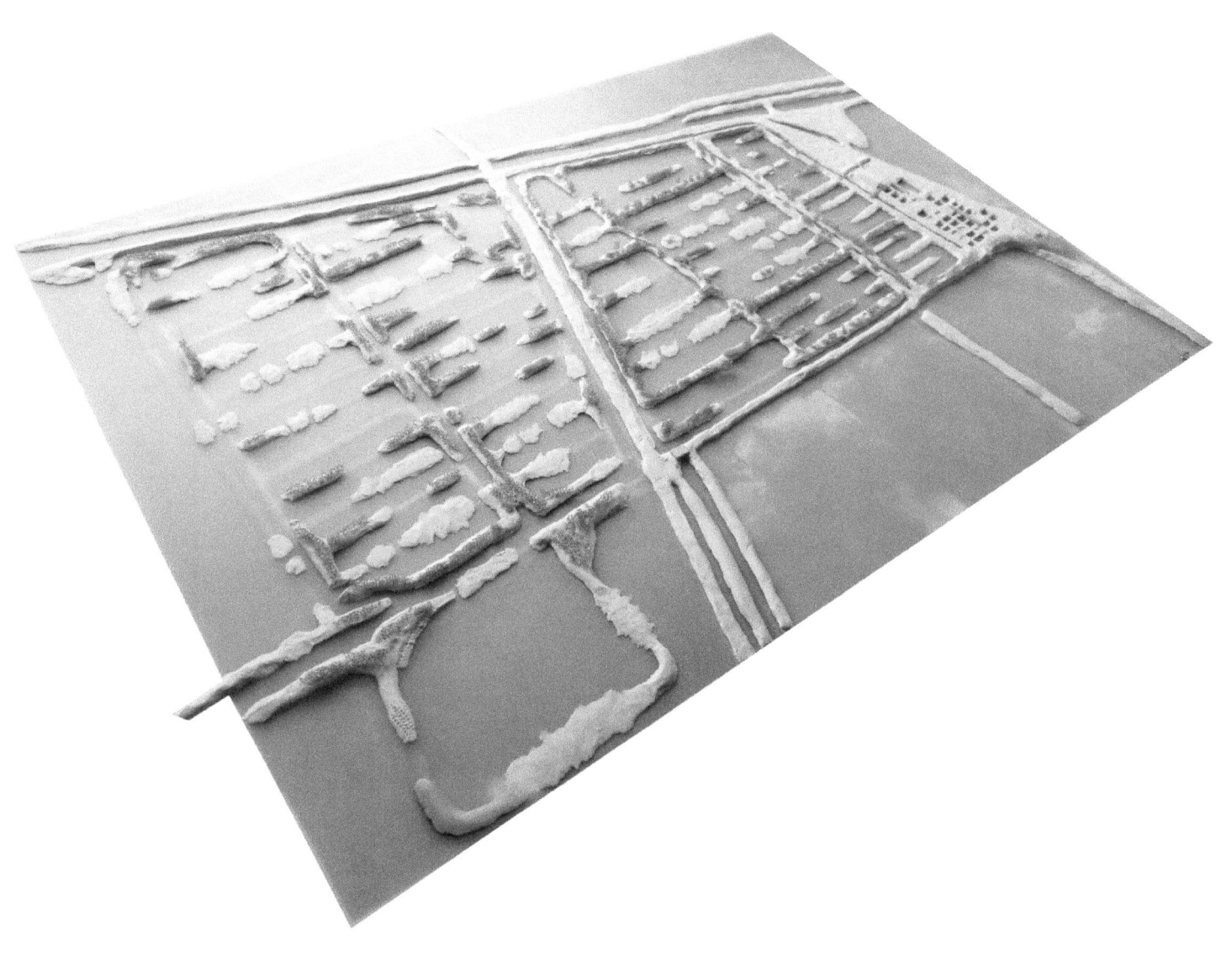

25 *La Tancada Lagoon Park*. EMF reshaped various dikes into four low profiles to create a range of contiguous habitat types, including Mediterranean salt steppes, mudflats, and coastal lagoons. The landscape is responsive to tidal flux, and a diversity of plants and animals flourish here, including the endemic and endangered Spanish Toothcarp (*Aphanius iberis*).

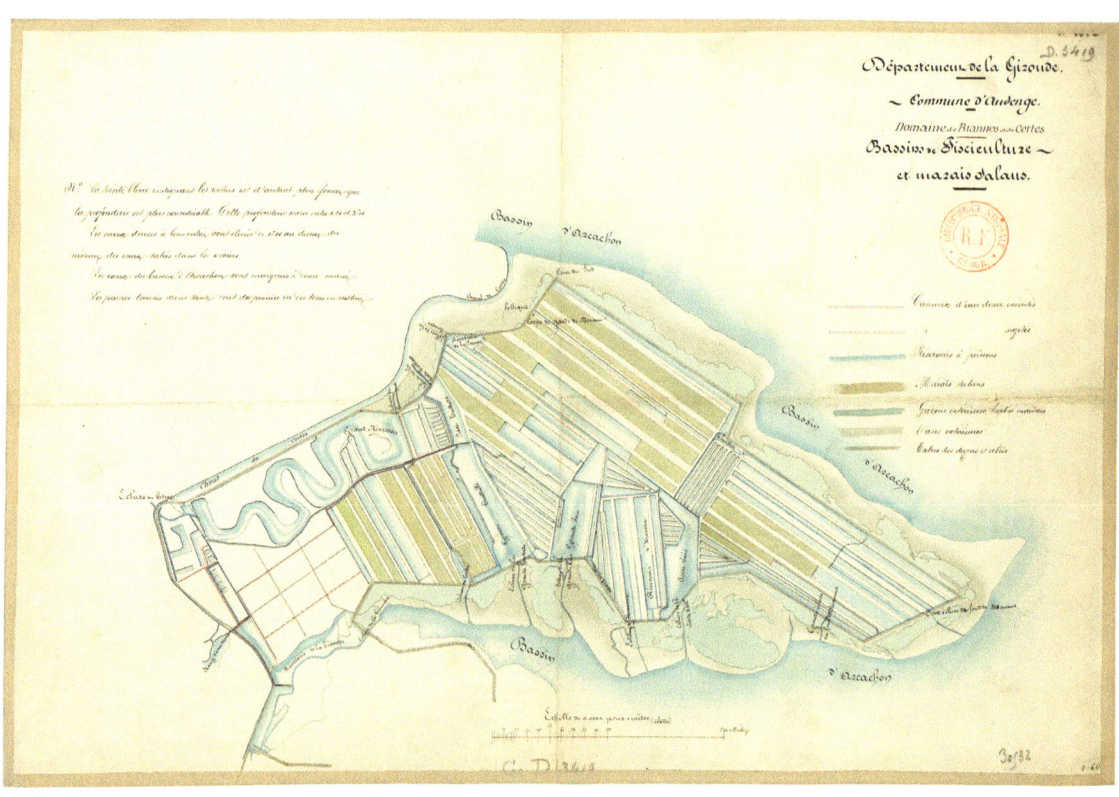

26 Artist unknown, *Bassins de pisciculture et marais salants*, nineteenth century. Plan of the Domaine de Certes at the Archachon Basin in the commune of Audenge, France. Originally built for salt production, this landscape was gradually adapted for aquaculture in the early nineteenth century. This plan depicts a time when salt and fish farming coexisted.

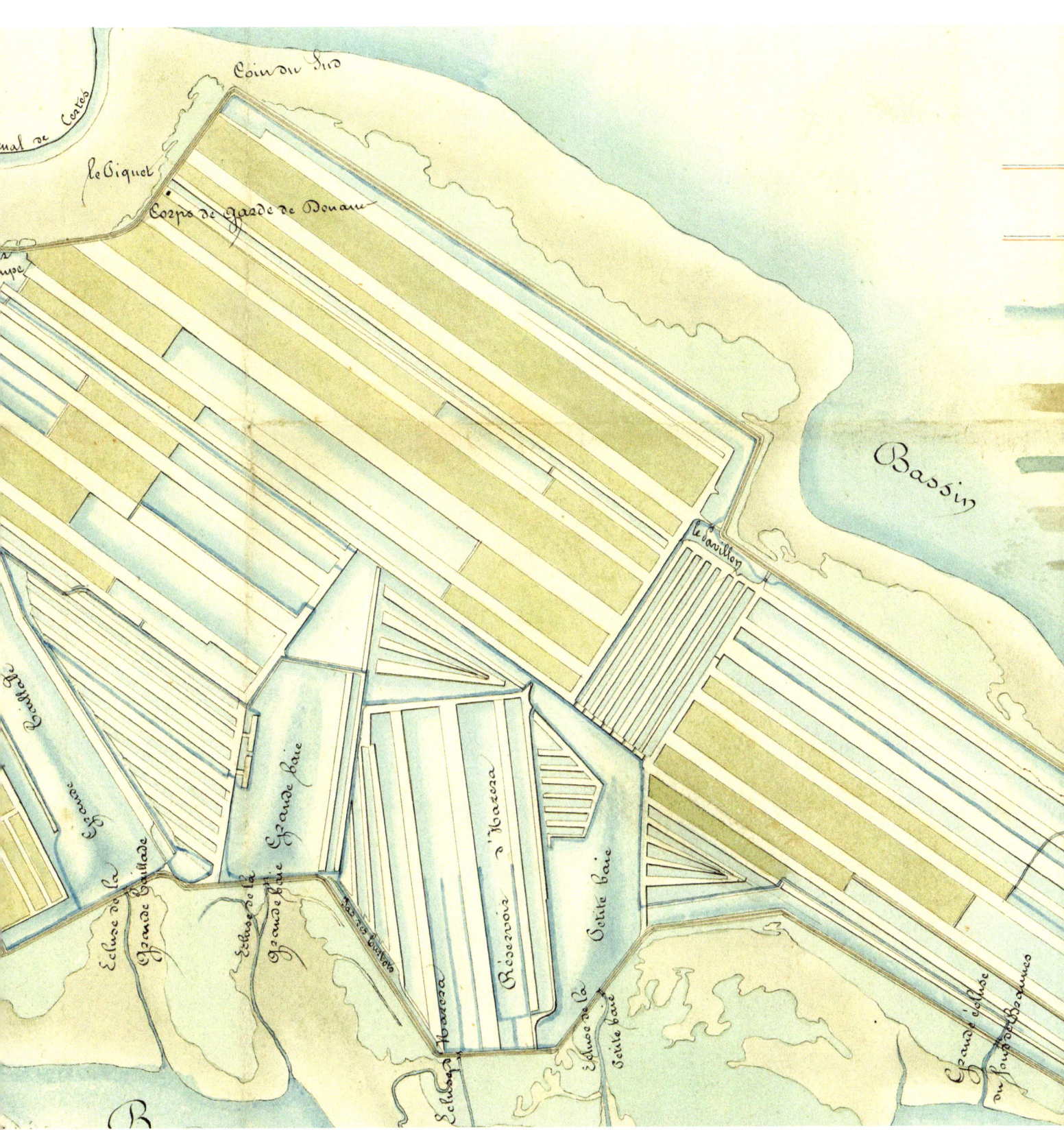

27 *Bassins de pisciculture et marais salants* (detail). Earthen dikes shaped basins for fish production (blue) and salt evaporation (green). The Domaine de Certes was a productive aquaculture, agricultural and forestry estate until the 1960s; today it is managed as a biodiverse nature reserve by the Conservatoire du Littoral and Gironde Departmental Council.

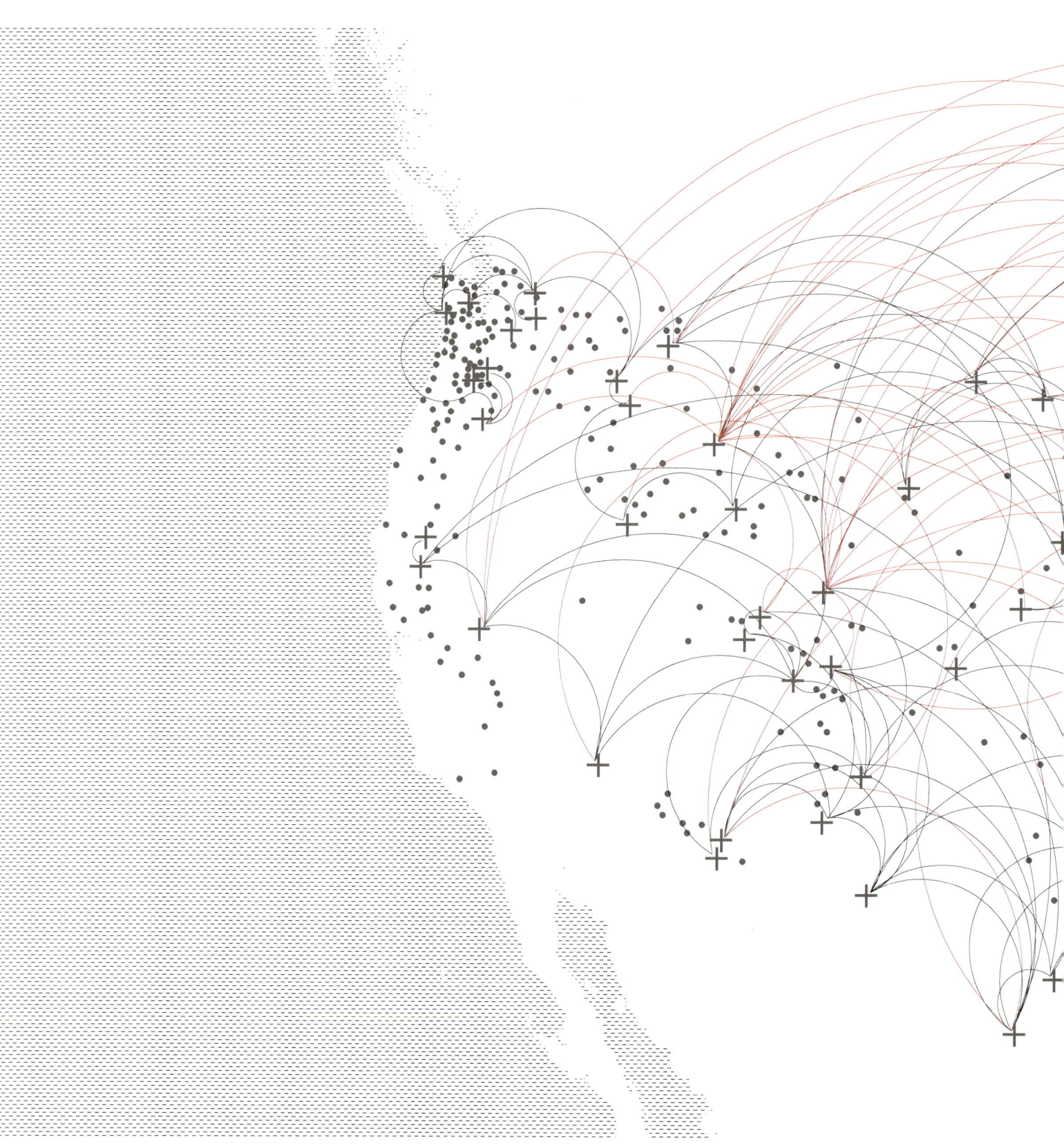

28 Michael Ezban, *Fish and Egg Exchanges*, digital composite, 2014. Map of US National and State fish hatcheries—a decentralized, continental-scale aquaculture-infrastructure system. Adapted from US Fish and Wildlife Service numeric data, the map spatializes the distribution of 121 million fish eggs and 164 million fish between national hatcheries in 1998.

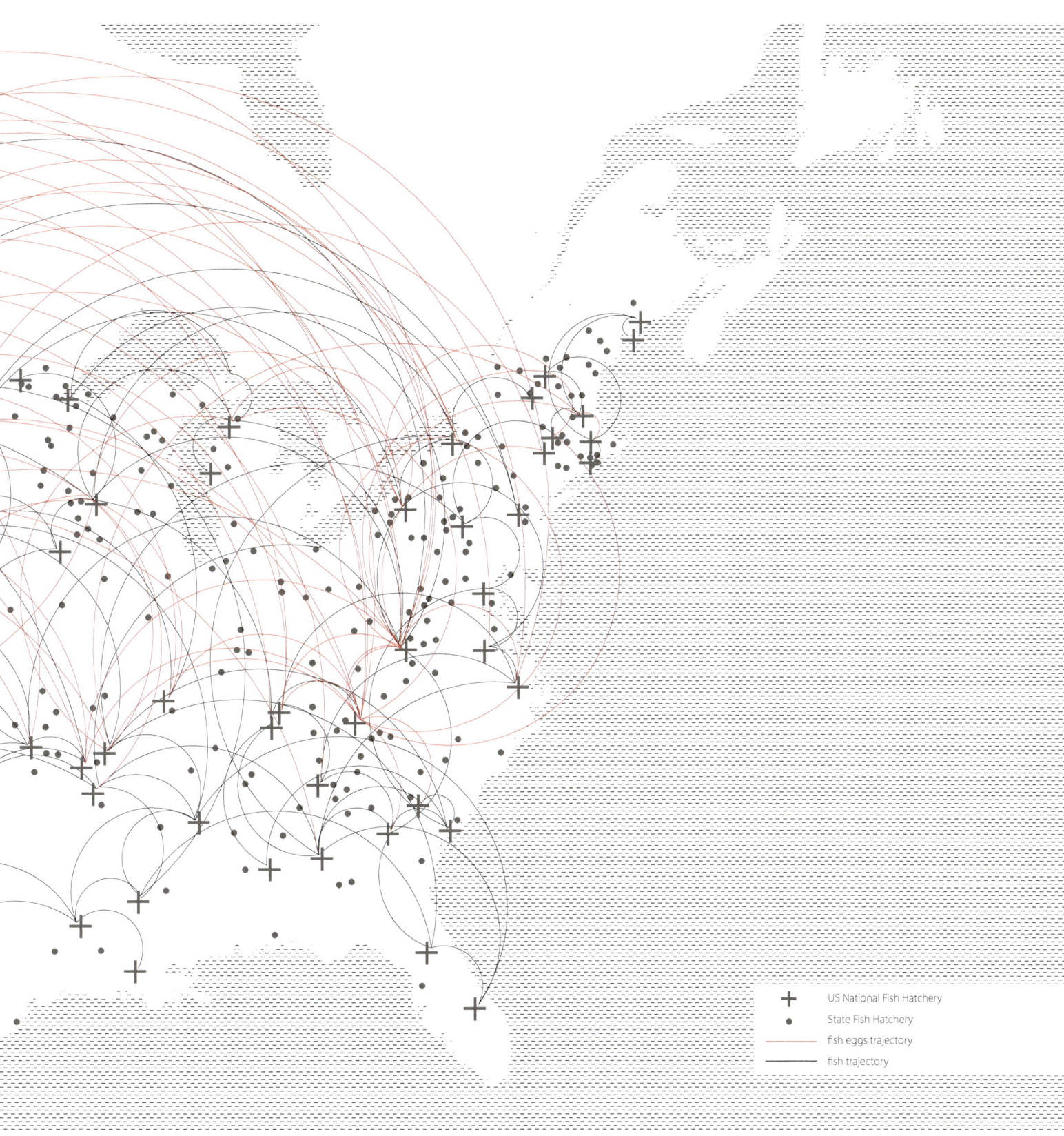

US National Fish Hatchery

State Fish Hatchery

fish eggs trajectory

fish trajectory

29 Artist unknown, sculpture at *Abri du Poisson, France,* ~25,000 BCE. The oldest known depiction of a fish by early humans is identified by experts as a spawning Atlantic Salmon *(Salmo salar)*. Carved into the ceiling of a shallow cave in the Dordogne, France, this life-size sculpture exemplifies the early human impulse to represent fish in limestone landscapes.

30 Sculpture at *Abri du Poisson* (detail). This detail depicts the sculpture at half of actual scale. Spawning salmon undergo significant skeletal alterations. In males, the most prominent change is the development of a kype (hook) in the tip of the lower jaw. The clear depiction of a kype on this salmon is key to identification of this fish as male.

01 David Benjamin and Natalie Jeremijenko, *Amphibious Architecture*, 2009. A field of high-tech buoys in the East River of New York City sense the presence of adjacent fish, and their light changes color relative to quantities of dissolved oxygen in the water. Dialogue between humans and fish is possible through an SMS interface.

Afterword

In his book *Being Salmon, Being Human*, ecological philosopher Martin Lee Mueller theorizes a relational constitution of human identity. He writes, "Our sense of who we are as humans is mirrored in our lived relationships with other creatures."[1] In contrast to anthropocentric paradigms that separate "us" from "them," Mueller casts salmon and humans as "not alone, but participants, shareholders, accomplices."[2] The reciprocal relations he describes are opportunities for us to better know ourselves and others we perceive to be unlike ourselves. In these more-than-human dialogues our identities are neither fixed or predetermined; we are *becoming*. And, as Donna Haraway reminds us, "becoming is always becoming *with*."[3] Entangled fish and humans produce "unpredictable kinds of 'we.'"[4]

Fish and humans do not only coshape each other; together, we coshape the landscapes in which we work, play, forage, and socialize. Our aquaculture landscapes enable a resilient, efficient, and biodiverse "zoöpolis," and they support "livelier livelihoods" for the allied fish and humans on the farm.[5] And, importantly, humans design aquaculture landscapes where embodied, sensorial experiences spark in us a sense of wonder for the remarkable ways of our "underwater cousins."[6] Elizabeth Meyer writes that aesthetic experiences in designed landscapes are transformative, they re-center human consciousness from an egocentric to a more bio-centric perspective.[7] She advocates the creation of landscapes as "exaggerated" versions of nature, employing design tactics such as *amplification*, *juxtaposition*, and *displacement*, to awaken and attune us to "previously unrealized relationships between human and non-human life processes."[8]

Two recent design projects illustrate such approaches. In 2009, architect David Benjamin and artist Natalie Jeremijenko deployed buoys in the East River of New York City that use sonar and hydrophones to detect fish and light up as fish swim by. The buoy lights shift color to register quantities of dissolved oxygen critical to fish survival. Humans partake in data-driven dialogues with the fish via an SMS interface and will soon be able to remotely release food to those with whom they are socializing.[9] This interface amplifies the typically invisible presence of fish and creates a novel bridge across the species divide.

In another example, the design team, Cooking Sections, built an "oyster table" within the tidal zone of Loch Portree, in Scotland. At high tide, oysters growing within the submerged metal-frame table filter and cleanse water. At low tide the table is a communal site used to facilitate urgent dialogue. In 2017, local politicians, residents, aquaculture stakeholders, and environmentalists sat together to discuss the ecological challenges posed by polluting types of fish farming.[10] Oysters literally had a seat at the table.

Tech-assisted fish-human dialogues and intertidal tables for multispecies coalitions augment the lineage of aquaculture landscapes that have, for millennia, enabled intricate choreographies of capture, culture, and cohabitation. In the twenty-first century, we partner with tilapia to manage urban wastestreams, align salmon migration corridors with public esplanades, integrate fish farming with habitat conservation and ecotourism, ally with oysters to protect our coastlines and filter our water, and imaginatively illustrate ichthyocentric perspectives. As we continue the ancient project of entwining our worlds, humans invite and renew the possibility of becoming with fish.

Notes

Foreword

1. British primatologist Richard Wrangham argues that we have been eating fish, among other forms of animal flesh, since long before we were human. Rather than *Homo sapiens* domesticating fire, Wrangham argues, the adaptation from pre-human hominids such as *Homo erectus* to *Homo sapiens* was fueled by the caloric and nutritional benefits of cooking. See Richard Wrangham, *Catching Fire: How Cooking Made Us Human* (New York: Basic Books, 2009).

Introduction

1. Edward Daeschler et al., "A Devonian Tetrapod-like Fish and the Evolution of the Tetrapod Body Plan," *Nature* 440 (April 2006): 757.
2. Ibid, 757.
3. Neil Shubin, *Your Inner Fish* (New York: Vintage Books 2009), 212.
4. Teresa E. Steele, "A Unique Hominin Menu Dated to 1.95 Million Years Ago," *Proceedings of the National Academy of Sciences* 107, no. 24 (June 2010): 10771-10772.
5. Curtis Marean, "Early Human Use of Marine Resources and Pigment in South Africa during the Middle Pleistocene," *Nature* 449 (October 18, 2007): 905.
6. "Abri du Poisson Cave," *Encyclopedia of Stone Age Art*, accessed March 1, 2019, http://www.visual-arts-cork.com/prehistoric/abri-poisson.html.
7. Ian Mcniven et al., "Phased Redevelopment of an Ancient Gunditjmara Fish Trap Over the Past 800 Years: Muldoons Trap Complex, Lake Condah, Southwestern Victoria," *Australian Archaeology* 18 (December 2015): 44.
8. Ibid, 44-45.
9. Anna Salleh, "Aborigines May Have Farmed Eels, Built Huts," *ABC Science*, accessed March 1, 2019, http://www.abc.net.au/science/articles/2003/03/13/806276.htm.
10. John Lucas et al., *Aquaculture: Farming Aquatic Animals and Plants*, 3rd ed. (Oxford: Wiley-Blackwell, 2012), 2.
11. Ibid, 3-5.
12. Martin Beveridge and David Little, "The History of Aquaculture in Traditional Societies," in *Ecological Aquaculture: The Evolution of the Blue Revolution*, ed. Barry Costa-Pierce (Oxford: Blackwell Science, 2002), 25-26.

13. M. Anders Halverson, "Stocking Trends: A Quantitative Review of Governmental Fish Stocking in the United States, 1931 to 2004," *Fisheries* 33, no. 2, (February 2008): 69.
14. Darin Kinsey, "Seeding the Water as the Earth: The Epicenter and Peripheries of a Western Aquaculture Revolution," *Environmental History* 11, no. 3 (July 2006): 535-536.
15. Ibid, 552.
16. Kevin Glover et al., "Half a Century of Genetic Interaction Between Farmed and Wild Atlantic Salmon: Status of Knowledge and Unanswered Questions," *Fish and Fisheries* 18 (2017): 891.
17. Harvey Neo, "Aquaculture," in *Humans and Animals: A Geography of Coexistence*, ed. Julie Urbanik and Connie Johnston (Santa Barbara: ABC-CLIO, 2017): 29-30.
18. "The Fish Feed Story," *The Fish Site*, accessed March 1, 2019, https://thefishsite.com/articles/the-fish-feed-story.
19. Peter Edwards, "Aquaculture Environment Interactions: Past Present and Likely Future Trends," *Aquaculture* 447 (February 2015): 3.
20. Glover et al., "Half a Century of Genetic Interaction Between Farmed and Wild Atlantic Salmon," 916.
21. Edwards, "Aquaculture Environment Interactions," 3-4.
22. United Nations Food and Agriculture Organization, *The State of World Fisheries and Aquaculture 2016 Contributing to Food Security and Nutrition for All* (Rome: FAO, 2016), 2.
23. Kate Orff, *Toward an Urban Ecology* (New York: Monacelli Press, 2016), 86.
24. Julie Urbanik, *Placing Animals: An Introduction to the Geography of Human-Animal Relations* (Lanham: Rowman & Littlefield Publishers, 2012), 3.
25. Jonathan Balcombe, *What a Fish Knows: The Inner Lives of Our Underwater Cousins* (New York: Scientific American/Farrar, Straus and Giroux, 2016), 6.
26. See Peter Singer, *The Expanding Circle* (Princeton: Princeton University Press, 1981).
27. Barry Costa-Pierce, ed., *Ecological Aquaculture: The Evolution of the Blue Revolution* (Oxford: Blackwell Science, 2002), xiii.
28. Doris Soto et al., eds., "Building an Ecosystem Approach to Aquaculture"

(Rome: FAO, 2008), 3-5.
29. Jody Emel et al., "Livelier Livelihoods: Animal and Human Collaboration on the Farm," in *Critical Animal Geographies: Politics, Intersections and Hierarchies in a Multispecies World*, ed. Kathryn Gillespie et al. (Taylor and Francis, 2015), 171.
30. For example, see Michael Ezban, "Decoys, Dikes, and Lures: Polyfunctional Landscapes of Waterfowl Hunting," *Studies in the History of Gardens and Designed Landscapes* 33, no. 3 (2013): 1-6.
31. Tom Williamson, *Polite Landscapes: Gardens and Society in Eighteenth-Century England* (Baltimore: Johns Hopkins University Press, 1995), 119.
32. Christopher Currie, "Fishponds as Garden Features, c. 1550-1750," *Garden History* 18, no. 1 (Spring, 1990): 24-25.
33. E. Dennison and R. Iles, "Medieval Fishponds in Avon," in *Bristol & Avon Archaeology 4*, (Bristol: Bristol and Avon Archaeological Society, 1985), 41.
34. Stephen Switzer, *Ichnographia Rustica: or, the Nobleman, Gentleman, and Gardener's Recreation, no. 3* (London: D. Browne, 1718), 113-127.
35. Currie, "Fishponds as Garden Features," 42.
36. See Eugene Odum, *Fundamentals of Ecology* (Philadelphia: W.B. Saunders Company, 1953).
37. See Charles Waldheim, *Landscape as Urbanism: A General Theory* (Princeton: Princeton Architectural Press, 2016); and James Corner, "Terra Fluxus," in *The Landscape Urbanism Reader*, ed. Charles Waldheim (Princeton: Princeton Architectural Press, 2006).
38. Orff, *Toward an Urban Ecology*, 81.
39. "Mutualism by Design," SCAPE, accessed Jan 25, 2019, https://www.scapestudio.com/ideas/.
40. Elizabeth K Meyer, "Beyond 'Sustaining Beauty': Musings on a Manifesto," in *Values in Landscape Architecture*, ed. Elen Deming (Baton Rouge: Louisiana State University, 2015), 47.
41. Ibid, 34.
42. Urbanik, *Placing Animals*, 8.
43. Jennifer Wolch, "Zoöpolis," in *Animal Geographies: Place Politics, and Identity in the Nature-Culture Borderlands*, ed. Jennifer Wolch and Jody Emel (London: Verso, 1998), 124.
44. Catherine Johnston, "Beyond the Clearing: Toward a Dwelt Animal

Geography," *Progress in Human Geography* 32, no. 5 (2008): 646.

45. Ibid, 646.

46. Emel et al., "Livelier Livelihoods," 178.

47. Ibid, 167.

48. Donna Haraway, *When Species Meet* (Minneapolis: University of Minnesota Press, 2007), 74.

49. Ibid, 80.

50. Christopher Bear and Sally Eden, "Thinking Like a Fish? Engaging through Nonhuman Difference through Recreational Angling," *Environment and Planning D* 29, no. 2 (2011): 349.

51. Ibid, 336.

52. Ibid, 349.

53. Haraway, *When Species Meet*, 296.

54. Ibid, 81.

55. Soto et al., "Building an Ecosystem Approach to Aquaculture," 3.

56. Cecile Brugere et al., "The Ecosystem Approach to Aquaculture 10 Years On," *Reviews in Aquaculture* 0 (2018): 12.

57. Alan Desbonnet and Barry Costa-Pierce, "Aquaculture in Future Urban Ecosystems," in *Urban Aquaculture*, ed. Costa-Pierce et al. (Oxfordshire: CABI, 2005), 277.

58. Ibid, x.

59. "Species by Family/Subfamily," *Eschmeyer's Catalog of Fishes*, accessed March 1, 2019, http://researcharchive. calacademy.org/research/ichthyology/catalog/SpeciesByFamily.asp.

60. United Nations Food and Agriculture Organization, *The State of World Fisheries and Aquaculture 2018* (Rome:FAO, 2018), 21.

Designing Ichthyological Urbanism

1. Orff, *Toward an Urban Ecology*, 89-107.

2. Colin Nash, *The History of Aquaculture* (Ames: Wiley-Blackwell, 2011), 26.

3. "Sunqiao Urban Agricultural District," *Sasaki*, accessed March 1, 2019, http://www.sasaki.com/project/417/sunqiao-urban-agricultural-district/.

4. Wolch, "Zoöpolis," 124.

5. Jennifer Wolch, "Anima Urbis," *Progress in Human Geography* 26, no. 2 (2002): 729.

6. Ibid, 729.

7. See Menno Schilthuizen, *Darwin Comes to Town: How the Urban Jungle Drives Evolution* (New York: Picador, 2018).

8. See Balcombe, *What a Fish Knows.*

9. Jeffrey Cordell et. al., "Benches, Beaches, and Bumps: How Habitat Monitoring and Experimental Science Can Inform Urban Seawall Design," in *Living Shorelines: The Science and Management of Nature-Based Coastal Protection*, ed. Donna Bilkovic et al. (Boca Raton: CRC Press, 2017), 422.

10. Nina-Marie Lister, "Resilience: Designing the New Sustainability," *Topos* 90, (2015): 20.

11. "Oyster Reef Restoration," *The Nature Conservancy*, accessed March 1, 2019, https://www.nature.org/en-us/about-us/where-we-work/united-states/south-carolina/stories-in-south-carolina/oyster-reef-restoration-southern-solutions-for-a-global-problem-1.

12. Antonio Rodríguez et al., "Oyster Reefs Can Outpace Sea-level Rise," *Nature Climate Change* 4, (June 2014): 493-497.

13. Orff, *Toward an Urban Ecology*, 237.

14. Christophe Girot, *The Course of Landscape Architecture* (London: Thames and Hudson, 2016), 327.

15. Brian Davis, "Public Sediment," in *Toward an Urban Ecology*, Kate Orff (New York: The Monacelli Press, 2016), 233-234.

16. Tom Leader Studio, "Making Ground/ Farming Water," *Ground Up Journal*, accessed March 1, 2019, http://groundupjournal.org/tom-leader/.

17. N.C. Nelson, "Shellmounds of the San Francisco Bay Region," *University of California Publications in American Archaeology and Ethnology* 7, no. 4 (1909): 325.

18. Ibid, 337.

19. Kent Lightfoot, "Shellmounds: An Archaeologist's View," *News from Native California* 17, no. 3 (Spring 2004): 16.

20. Ibid, 16.

21. N.C. Nelson, "Shellmounds of the San Francisco Bay Region," 348.

22. Tom Leader, "Sediment," *Landscape Journal* 21, no. 2 (2002): 75-76; and Mathur/Da Cunha and Tom Leader Studio, "DYNAMIC COALITION," *PRAXIS: Journal of Writing + Building*, no. 4 (2002): 40-47.

23. Philipe Coignet, "Foreword" in *Tom Leader Studio: Three Projects*, ed. Jason Kentner (Princeton: Princeton Architectural Press, 2010), 7.

24. James Corner Field Operations et. al., "Design Summary: Concept Design and Framework Plan for Seattle's Central Waterfront," (July 2012), 52.

25. Jeffrey Cordell et. al., "Benches, Beaches, and Bumps," 423.

26. "Chinook Salmon—Protected," *NOAA Fisheries*, accessed March 1, 2019, https://www.fisheries.noaa.gov/species/chinook-salmon-protected.

27. Dale Stokes, *The Fish in the Forest: Salmon and the Web of Life* (Oakland: University of California Press, 2014), 125.

28. Hatchery Science Review Group, Report to Congress on the Science of Hatcheries (June 2015), Introduction.

29. Ibid, Introduction.

30. Ibid, 3.

31. Cordell et. al., "Benches, Beaches, and Bumps," 423.

32. "Seawall Strata," *Haddad/Drugan*, accessed March 1, 2019, http://www.haddad-drugan.com/#/seawall-strata/.

33. Haddad|Drugan, Elliot Bay Seawall Project Art Programming Plan, (January 2013), 25.

34. "Olympic Sculpture Park," The Landscape Architecture Foundation, accessed March 1, 2019, https://www.landscapeperformance.org/case-study-briefs/olympic-sculpture-park.

35. Paul Greenberg, *Four Fish: The Future of the Last Wild Food* (New York: Penguin, 2010), 255.

36. "The Seawater Energy and Agriculture System," *Sustainable Bioenergy Research Consortium*, accessed March 1, 2019, https://sbrc.masdar.ac.ae/index.php/projects/seas/item/76-the-seawater-energy-and-agriculture-system.

37. John Matthews, "Could a Solution to Reducing Aviation Emissions be Found in the Arabian desert?" *CNN World*, accessed March 1, 2019, https://www.cnn.com/2017/07/12/middleeast/iseas-abu-dhabi-aviation-biofuel/index.html.

38. See Sonja Duempelmann and Charles Waldheim, eds., *Airport Landscape: Urban Ecologies in the Aerial Age* (Cambridge: Harvard University Graduate School of Design, 2016).

39. Elizabeth Royte, "Street Farmer," *New York Times Magazine*, last modified July 1, 2009, https://www.nytimes.com/2009/07/05/magazine/05allen-t.html.

40. Herbert Wright, "Floating Fields Wins Shenzhen UABB Award and is Set to Continue through 2016," *ArchDaily*, last modified March 17, 2016, https://www.archdaily.com/783314/floating-fields-wins-shenzhen-uabb-award-and-is-set-to-continue-through-20169.

41. Sharon Haar, "Heritage and Sustainability in Shunde (China)," in *On Location: Heritage Cities and Sites*, ed. D. Fairchild Ruggles (New York: Springer, 2012), 228-229.

42. Nina-Marie Lister, "Sustainable Large Parks: Ecological Design or Designer Ecology?" in *Large Parks*, ed. George Hargreaves and Julia Czerniak (Princeton: Princeton Architectural Press, 2007), 35.

43. Margaret Crawford, "Urban Agriculture in the Pearl River Delta," in *Food and the City: Histories of Culture and Cultivation*, ed. Dorothée Imbert (Washington DC: Dumbarton Oaks, 2016), 251.

44. Ibid, 267.

45. Pat Dorsey, *Fly Fishing Tailwaters: Tactics and Patterns for Year-round Waters* (Mechanicsburg: Stackpole, 2009), 6-14.

46. Anders Halverson, *An Entirely Synthetic Fish: How Rainbow Trout Beguiled America and Overran the World* (New Haven: Yale UP, 2010), 186.

47. See for example US Department of Agriculture National Agricultural Statistics Service, *2013 Census of Aquaculture*, vol. 3, September 2014; and "Agritourism Program," *Massachusetts Department of Agricultural Resources*, accessed March 1, 2019, https://www.mass.gov/agritourism-program.

48. Anne Whiston Spirn, "Constructing Nature: The Legacy of Frederick Law Olmsted," in *Uncommon Ground: Rethinking the Human Place in Nature*, ed. William Cronon (New York: Norton & Company, 1996), 113.

49. Sebastien Marot, "The Reclaiming of Sites," in *Recovering Landscape: Essays on Contemporary Landscape Theory*, ed. James Corner (Princeton: Princeton Architectural Press, 1999), 50.

50. Susan Herrington, *Landscape Theory in Design* (New York: Routledge, 2017), 133.

51. James Corner, "Three Tyrannies of Contemporary Theory," in *The Landscape Imaginary*, ed. James Corner and Alison Hirsch (Princeton: Princeton Architectural Press, 2014), 101.

52. Ibid, 101.

53. See Richard C. Hoffman, "Economic Development and Aquatic Ecosystems in Medieval Europe," *American Historical Reviews* 101 (1996): 630-669.

54. Hansjörg Gadient, "Specific Landscapes," in *Specific Landscapes*, hr+c Landschaftsarchitektur (Berlin: Bovis, 2012), 13.

55. Ibid, 19.

56. John Dixon Hunt, "Is Landscape History?" in *Is Landscape…? Essays on the Identity of Landscape*, ed. Gareth Doherty and Charles Waldheim, (London: Routledge, 2016), 253.

57. hr+c Landschaftsarchitektur, *Specific Landscapes*, 177.

58. Herrington, *Landscape Theory in Design*, 160.

59. See for example Faith Chan, et al., "'Sponge City' in China—A Breakthrough of Planning and Flood Risk Management in the Urban Context," *Land Use Policy* 76 (2018): 772-778.

60. See Elizabeth Meyer, "Sustaining Beauty: The Performance of Appearance," *Journal of Landscape Architecture* (Spring 2008).

61. Kongjian Yu, "The Big Foot Revolution," in *Designed Ecologies: The Landscape Ecology of Kongjian Yu*, ed. William S. Saunders (Basel: Birkhäuser, 2012), 42-43.

62. Ibid, 43.

63. Catherine Seavitt, "Yangtze River Delta Project," *Scenario Journal* 03 (Spring 2013), accessed March 1, 2019, https://scenariojournal.com/article/yangtze-river-delta-project/.

64. WWF, "Yangtze Freshwater Aquaculture," accessed March 1, 2019, https://www.wwf.org.uk/sites/default/files/2017-04/161128_Yangtze_Aquaculture_CS_Final.pdf.

65. Kelly Shannon, "(R)evolutionary Ecological Infrastructures," in *Designed Ecologies: The Landscape Ecology of Kongjian Yu*, ed. William S. Saunders, (Basel: Birkhäuser, 2012), 200.

66. "Yichang Yunhe Park," *Turenscape*, accessed March 1, 2019, https://www.turenscape.com/en/project/detail/4638.html.

Case Study 01

1. International Union for Conservation of Nature and Natural Resources, *Fishing for a Living: Ecology and Economics of Fishponds in Central Europe* (Cambridge: ICUN, 1997), 2.

2. J. Květ and Jan Jeník, eds., *Freshwater Wetlands and Their Sustainable Future: A Case Study of the Trebon Basin Biosphere Reserve* (Boca Raton: Parthenon Pub. Group, 2002): xviii.

3. International Union for Conservation of Nature and Natural Resources, *Fishing for a Living*, 16.

4. Richard Lhotsky, "The Role of Historical Fishpond Systems During Recent Flood Events," *Journal of Water and Land Development* 14, no. 1 (2012): 53.

Case Study 02

1. Yufera M. and A.M. Arias, "Traditional Polyculture in Esteros in the Bay of Cadiz," *Aquaculture Europe* 35, no. 3 (September 2010): 22.

2. Ibid., 23.

3. Tanja Michler, "Management of Marine Bio-Resources: A Comparative Study of Three European Coastal Areas," (master thesis, Carl Von Ossietzky Universität, 2003): 28.

Case Study 03

1. Miguel Medialdea, "A New Approach to Sustainable Aquaculture," *Solutions Journal* 1, no. 3 (May 2010): 14.

2. Ibid,14.

3. Lisa Abend, "Sustainable Aquaculture: Net Profits," *Time Magazine*, June 15, 2009.

4. United Nations Environment Program, "Ecosystem Approach to Aquaculture Management and Biodiversity Conservation in a Mediterranean Coastal Wetland," (May 24, 2012): 11-12.

Case Study 04

1. S. Cataudella et al., "Mediterranean Coastal Lagoons," in *Studies and Reviews General Fisheries Commission for the Mediterranean*, no. 95 (Rome: FAO, 2015), 117.

2. Nash, *The History of Aquaculture*, 27.

3. Folco Cecchini, *Sorella Anguilla: Pesca e manifattura nelle valli di Comacchio* (Bologna: Minerva Edizioni, 2011), 57.

4. Folco Cecchini, ed., *The Comacchio Lagoon: A Naturalistic and Historical Itinerary* (Bologna: Tipografia Compositori, 1989), 22.

Case Study 05

1. James Higginbotham, *Piscinae: Artificial Fishponds in Roman Italy* (Chapel Hill: The University of North Carolina Press, 1997), 31-32.

2. Ibid, 72.

3. Ibid, 27.

4. A. F. Stewart, "To Entertain an Emperor: Sperlonga, Laokoon, and Tiberius at the Dinner-Table," *The Journal of Roman Studies* 67 (1977): 76-90.

Case Study 06

1. Nash, *The History of Aquaculture*, 57.

2. Darin Kinsey, "Seeding the Water as the Earth," 535-536.

3. James Bertram, *The Harvest of the Sea* (London: Alexander Gardner, 1885), 84.

4. W.H. Fry, *A Complete Treatise on Artificial Fish Breeding* (New York: D. Appleton and Company, 1854), 103.

5. Nash, *The History of Aquaculture*, 57.

6. Thomas Ashworth, "French Pisciculture Establishment at Huningue, near Basle," *The Field, the Country Gentleman's Newspaper*, August 25, 1860, 169.

Case Study 07

1. Thomas Ferguson, "Pisciculture," in *Reports of the United States Commissioners to the Paris Universal Exposition 1878*, no. 5 (Washington: Government Printing Office, 1880), 449.

2. Ibid, 450.

3. K. Ann, "Concrete and the Engineered Picturesque the Parc des Buttes Chaumont (Paris, 1867)," *Journal of Architectural Education* 58, no. 1 (September 2004): 5-12.

4. Paris (France) Aquarium du Trocadéro, *Notice Sur L'aquarium Du Trocadéro En 1900* (Paris: Imprimerie Nouvelle (Association Ouvrière), 1900), 34-36.

Case Study 08

1. James Kapetsky, *Some Considerations for the Management of Coastal Lagoons and Estuarine Fisheries* (Rome: FAO, 1981), 57.

2. R.L. Welcomme, "Traditional Brush Park Fisheries in Natural Waters," in *Periphyton: Ecology, Exploitation, and Management* (Oxfordshire: CABI Publishing, 2005), 153-154.

3. Ibid, 146.

4. Ibid, 152-153.

5. Gérard Chouquer, "La dynamique morphologique du foncier piscicole et agricole sur les rives du Lac Nokoué au Bénin," *ArcheoGeographie*, accessed March 1, 2019, http://www.archeogeographie.org/index.php?rub=dossiers/etudes/nokoue.

Case Study 09

1. Dhrubajyoti Ghosh, *Ecology and Traditional Wetland Practice: Lessons from Wastewater Utilisation in the East Calcutta Wetlands* (Kolkata: Worldview, 2005), 43.

2. Ibid, 57.

3. Ritesh Kumar and Nitai Kundu, eds., *East Kolkata Wetlands* (Kolkata: Wetlands International-South Asia, 2010), 10.

4. Ibid, 2.

5. Stephanie Carlisle, "Productive Filtration: Living System Infrastructure in Calcutta," *Scenario Journal* 03, (Spring 2013), accessed March 1, 2019, https://scenariojournal.com/article/productive-filtration/.

Case Study 10

1. Kenneth Ruddle and Gongfu Zhong, *Integrated Agriculture-aquaculture in South China: The Dike-pond System of the Zhujiang Delta* (Cambridge: Cambridge University Press, 1988), 9.

2. Ibid, 2.

3. Ibid, 16.

4. Peter Edwards, "Rural Aquaculture: From Integrated Carp Polyculture to Intensive Monoculture in the Pearl River Delta, South China," *Aquaculture Asia Magazine* (April-June 2008): 4.

Case Study 11

1. Soto, et al., "Building an Ecosystem Approach to Aquaculture," 177-179.

2. Rob Fletcher, "The Time is Ripe for Rice-Fish Culture," *The Fish Site*, last modified December 14, 2018, https://thefishsite.com/articles/the-time-is-ripe-for-rice-fish-culture.

3. Jian Xie, et al., "Ecological Mechanisms Underlying the Sustainability of the Agricultural Heritage Rice-Fish Coculture System," *Proceedings of the National Academy of Sciences* 88, no. 50 (December 2011): 1.

4. Yehong Sun, et al. "Tourism Potential of Agricultural Heritage Systems," *Tourism Geographies* 13, no. 1 (2011): 122.

Case Study 12

1. Richard Irving and Brian Morgan, *A Geography of the Mai Po Marshes* (Hong Kong: Hong Kong University Press, 1988), 27-33.

2. Austin Ramzy, "A Rural Patch of Hong Kong Where Rare Birds Sing and Developers Circle," *New York Times*, November 17, 2018, accessed March 1, 2019, https://www.nytimes.com/2018/11/17/world/asia/hong-kong-wetlands-mai-po-nam-sang-wai.html.

3. WWF Hong Kong, "Mai Po Wetland Habitat Fact Sheet: Gei wai," accessed March 1, 2019, http://awsassets.wwfhk.panda.org/downloads/gei_wai.pdf.

4. WWF Hong Kong, *Management Plan for the Mai Po Nature Reserve 2006-2010* (Hong Kong: WWF Hong Kong, 2006), 7-9.

Case Study 13

1. Clark Erickson, "An Artificial Landscape-scale Fishery in the Bolivian Amazon," *Nature* 409 (November 2000): 193.

2. Ibid, 191.

3. Ibid, 191.

4. Clark Erickson, "The Domesticated Landscapes of the Bolivian Amazon," in *Time and Complexity in Historical Ecology*, ed. William Balee and Clark Erickson (New York: Columbia University Press, 2006), 259.

5. Ibid, 263.

6. Charles Mann, "Earthmovers of the Amazon," *Science* 287 (Dec 2000): 789.

Case Study 14

1. Dorsey, *Fly Fishing Tailwaters*, 6.

2. US Fish and Wildlife Service, Wolf Creek National Fish Hatchery, (June 2017), 4-6.

3. George Athanasakes, "If You Build it, They Will Come: Designing a Self-sustaining Trout Stream," *Stantec*, last modified April 28, 2016, https://ideas.stantec.com/blog/if-you-build-it-they-will-come-designing-a-self-sustaining-trout-stream.

4. Kentucky Department of Fish and Wildlife Resources, "Mitigation Plan for Hatchery Creek," (2014), 15-18.

Case Study 15

1. Barry Costa-Pierce, "Aquaculture in Ancient Hawaii," *Bioscience* 37, no. 5 (May 1987): 325.

2. Ibid, 325.

3. Carol Wybon, *Tide and Current: Fishponds of Hawaii* (Honolulu: University of Hawaii Press, 1992), 104-107.

4. Dieter Mueller-Dombois and Nengah Wirawan, "The Kahana Valley Ahupua'a, a PABITRA Study Site on O'ahu, Hawaiian Islands," *Pacific Science* 59, no. 2 (2005): 308-311.

Representations of Aquaculture Landscapes

1. Encyclopedia of Stone Age Art, "Abri du Poisson Cave," accessed March 1, 2019, http://www.visual-arts-cork.com/prehistoric/abri-poisson.htm.

2. Ruediger Berghahn and Floris Bennema, "Ancient History of Flatfish Research," *Journal of Sea Research* 75 (2013): 3-7.

3. John Berger, "Past Present," The Guardian, October 12, 2002, accessed March 1, 2019, https://www.theguardian.com/artanddesign/2002/oct/12/art.artsfeatures3.

4. Beveridge and Little, "The History of Aquaculture in Traditional Societies," 8-9.

5. Cai Rekkui, et al., "Rice-Fish Culture in China: The Past, Present, and Future," in *Rice-Fish Culture in China*, ed. Kenneth MacKay (Ottawa: International Development Research Center, 1995), 3.

6. R. Günther, "The Oyster Culture of the Ancient Romans," *Journal of the Marine Biological Association of the United Kingdom* 4, no. 4 (1897): 360-365.

7. See Sir Robert Atkyns, *The Ancient and Present State of Gloucestershire* (London: W. Boyer, 1712).

8. See Yoshio Hiyama, *Gyotaku: The Art and Technique of the Japanese Fish Print* (Seattle: University of Washington Press, 1964).

9. Darin Kinsey, "Seeding the Water as the Earth," 542-543.

10. James Corner, "Representation and Landscape: Drawing and Making in the Landscape Medium," *Word & Image* 8, no. 3 (July-September 1992): 243.

Afterword

1. Martin Lee Mueller, *Being Salmon, Being Human: Encountering the Wild in Us and Us in the Wild* (London: Chelsea Green Publishing, 2017), xvi.

2. Ibid, xx.

3. Haraway, *When Species Meet*, 244.

4. Ibid, 5.

5. See Wolch, "Zoöpolis"; Emil et al., "Livelier Livelihoods."

6. See Balcombe, *What a Fish Knows*.

7. Meyer, "Sustaining Beauty," 6.

8. Ibid, 17-18.

9. Karen M'Closkey and Keith VanDerSys, *Dynamic Patterns: Visualizing Landscapes in a Digital Age* (New York: Routledge, 2017), 96-99.

10. "CLIMAVORE: On Tidal Zones," *Cooking Sections*, accessed March 1, 2019, http://www.cooking-sections.com/CLIMAVORE-On-Tidal-Zones.

Image Credits

Front and Back Cover
Design by Siena Scarff Design. Image courtesy of Bibliothèque nationale de France.

Table of Contents
FIG. 01 Courtesy of Pack-Shot/ Shutterstock.com

Foreword
FIG. 01 Courtesy of RMN-Grand Palais/ Martine Beck-Coppola photo.

Introduction
FIG. 01 By Michael Ezban.
FIG. 02 By Michael Ezban.

Part 01
By Michael Ezban.

Designing Ichthyological Urbanism
FIG. 01 Courtesy of SCAPE.
FIG. 02 Courtesy of Sasaki.
FIG. 03 Courtesy of Sasaki.
FIG. 04 Courtesy of Tom Leader/ TLS.
FIG. 05 Courtesy of Tom Leader/ TLS.
FIG. 06 Courtesy of Tom Leader/ TLS.
FIG. 07 Courtesy of Tom Leader/ TLS.
FIG. 08 Courtesy of Tom Leader/ TLS.
FIG. 09 Courtesy of James Corner/ Field Operations.
FIG. 10 Courtesy of James Corner/ Field Operations.
FIG. 11 Courtesy of Haddad|Drugan.
FIG. 12 Courtesy of Haddad|Drugan.
FIG. 13 Courtesy of Haddad|Drugan.
FIG. 14 Courtesy of Thomas Chung, School of Architecture, the Chinese University of Hong Kong.
FIG. 15 Courtesy of Thomas Chung.
FIG. 16 Courtesy of Thomas Chung.
FIG. 17 Courtesy of Thomas Chung.
FIG. 18 By Michael Ezban
FIG. 19 By Michael Ezban
FIG. 20 By Michael Ezban.
FIG. 21 By Michael Ezban.
FIG. 22 Courtesy of Christo Libuda (Lichtschwärmer)
FIG. 23 Courtesy of hutterreimann + cejka Landschaftsarchitektur.
FIG. 24 Courtesy of Christo Libuda (Lichtschwärmer)
FIG. 25 Courtesy of Jörg Hempel Fotografie.
FIG. 26 Courtesy of Google, © 2018 DigitalGlobe.
FIG. 27 Courtesy of Kongjian Yu/ Turenscape.

FIG. 28 Courtesy of Boyu Li.
FIG. 29 Courtesy of Kongjian Yu/ Turenscape.

Part 02
By Michael Ezban.

Overview of *Aquaculture Landscapes* Case Studies
FIG. 01 By Michael Ezban.
FIG. 02 By Michael Ezban.
FIG. 03 By Michael Ezban and David Bayer.
FIG. 04 By Michael Ezban.

Case Study 01
FIG. 01 By Michael Ezban.
FIG. 02 By Michael Ezban.
FIG. 03 By Michael Ezban.
FIG. 04 By Michael Ezban and David Bayer.
FIG. 05 By Michael Ezban.
FIG. 06 By Michael Ezban.
FIG. 07 By Michael Ezban.
FIG. 08 By Michael Ezban.
FIG. 09 By and courtesy of Andrew Klaver.
FIG. 10 By Michael Ezban.

Case Study 02
FIG. 01 By Michael Ezban.
FIG. 02 By Michael Ezban.
FIG. 03 By Michael Ezban.
FIG. 04 By Michael Ezban and David Bayer.
FIG. 05 Courtesy of Inmaculada Salado Reyes.
FIG. 06 By Michael Ezban.
FIG. 07 By Michael Ezban.
FIG. 08 By Michael Ezban.
FIG. 09 Courtesy of Inmaculada Salado Reyes.

Case Study 03
FIG. 01 By Michael Ezban.
FIG. 02 By Michael Ezban.
FIG. 03 By Michael Ezban.
FIG. 04 By Michael Ezban and David Bayer.
FIG. 05 Courtesy of Miguel Medialdea.
FIG. 06 Courtesy of Miguel Medialdea.
FIG. 07 By Michael Ezban.
FIG. 08 By Michael Ezban.
FIG. 09 By Michael Ezban.
FIG. 10 By Michael Ezban.

Case Study 04
FIG. 01 By and courtesy of Massimo Merlini.
FIG. 02 By Michael Ezban.
FIG. 03 By Michael Ezban.
FIG. 04 By Michael Ezban and David Bayer.
FIG. 05 By Michael Ezban.

FIG. 06 By Michael Ezban.
FIG. 07 By Michael Ezban.
FIG. 08 By Michael Ezban.
FIG. 09 By Michael Ezban.
FIG. 10 By Michael Ezban.
FIG. 11 By and courtesy of Nicola Armari.

Case Study 05
FIG. 01 By Michael Ezban.
FIG. 02 By Michael Ezban.
FIG. 03 By Michael Ezban.
FIG. 04 By Michael Ezban and David Bayer.
FIG. 05 By Michael Ezban.
FIG. 06 By Michael Ezban.
FIG. 07 By Michael Ezban.
FIG. 08 By Michael Ezban.
FIG. 09 By Michael Ezban.
FIG. 10 By Michael Ezban

Case Study 06
FIG. 01 Courtesy of Bibliothèque nationale de France. By Paul Langlois. Image from Charles Grad, "L'Alsace: le pays et ses habitants," *Le Tour du Monde* LXX (July 1887): 269.
FIG. 02 By Michael Ezban.
FIG. 03 By Michael Ezban.
FIG. 04 By Michael Ezban and David Bayer.
FIG. 05 © Illustrated London News Ltd/ Mary Evans. Image from "The Huningue Fish Nurseries," *The Illustrated London News*, February 10, 1861, 193.
FIG. 06 By Michael Ezban.
FIG. 07 By Michael Ezban.
FIG. 08 By Michael Ezban.
FIG. 09 Courtesy of Bibliothèque nationale de France. By Adolphe Braun. Image from *Etablissement de Pisciculture de Huningue (Haut-Rhin) Vues photographiques* (Paris: Department des Ponts et Chaussees, 1861).
FIG. 10 By Michael Ezban.

Case Study 07
FIG. 01 Courtesy of L'Illustration. Image from *L'Illustration Journal Universel* LXXI, no. 1842 (1878): 392.
FIG. 02 By Michael Ezban.
FIG. 03 By Michael Ezban.
FIG. 04 By Michael Ezban and David Bayer.
FIG. 05 Courtesy of Bibliothèque nationale de France. Image from *L'Art et L'Industrie de Tous les Peuples a L'Exposition Universelle de 1878* (Paris: Librairie Illustree, 1878), 17.
FIG. 06 Courtesy of Bibliothèque nationale de France. By Agence de presse Meurisse.

FIG. 07 By Michael Ezban.
FIG. 08 By Michael Ezban.
FIG. 09 Courtesy of Bibliothèque nationale de France. By Agence de presse Meurisse.

Case Study 08
FIG. 01 By and courtesy of Iwan Baan.
FIG. 02 By Michael Ezban.
FIG. 03 By Michael Ezban.
FIG. 04 By Michael Ezban and David Bayer.
FIG. 05 By and courtesy of Iwan Baan.
FIG. 06 By Michael Ezban.
FIG. 07 By Michael Ezban.
FIG. 08 By and courtesy of Iwan Baan.
FIG. 09 By and courtesy of Brian McMorrow.

Case Study 09
FIG. 01 By and courtesy of Brett Cole.
FIG. 02 By Michael Ezban.
FIG. 03 By Michael Ezban.
FIG. 04 By Michael Ezban and David Bayer.
FIG. 05 By and courtesy of Brett Cole.
FIG. 06 By and courtesy of Peter Edwards.
FIG. 07 By Michael Ezban.
FIG. 08 By Michael Ezban.
FIG. 09 By and courtesy of Brett Cole.
FIG. 10 By and courtesy of Peter Edwards.

Case Study 10
FIG. 01 Courtesy of Google, © 2018 DigitalGlobe.
FIG. 02 By Michael Ezban.
FIG. 03 By Michael Ezban.
FIG. 04 By Michael Ezban and David Bayer.
FIG. 05 Courtesy of Kenneth Ruddle.
FIG. 06 Courtesy of Kenneth Ruddle.
FIG. 07 By Michael Ezban.
FIG. 08 By Michael Ezban.
FIG. 09 Courtesy of Kenneth Ruddle.
FIG. 10 Courtesy of Google, © 2018 DigitalGlobe.

Case Study 11
FIG. 01 Courtesy of Liang Luohui.
FIG. 02 By Michael Ezban.
FIG. 03 By Michael Ezban.
FIG. 04 By Michael Ezban and David Bayer.
FIG. 05 Courtesy of Liang Luohui.
FIG. 06 Courtesy of Liang Luohui.
FIG. 07 By Michael Ezban.
FIG. 08 By Michael Ezban.
FIG. 09 Courtesy of Liang Luohui.
FIG. 10 Courtesy of Liang Luohui.
FIG. 11 Courtesy of Liang Luohui.

Case Study 12
FIG. 01 By and courtesy of Martin Harvey.
FIG. 02 By Michael Ezban.
FIG. 03 By Michael Ezban.
FIG. 04 By Michael Ezban and David Bayer.
FIG. 05 By and courtesy of Neil Fifer.
FIG. 06 Courtesy of Google, © 2018 DigitalGlobe.
FIG. 07 By Michael Ezban.
FIG. 08 By Michael Ezban.
FIG. 09 By and courtesy of Neil Fifer.
FIG. 10 By and courtesy of Gary Suen.
FIG. 11 By and courtesy of Gary Suen.

Case Study 13
FIG. 01 Courtesy of Google, © 2018 DigitalGlobe.
FIG. 02 By Michael Ezban.
FIG. 03 By Michael Ezban.
FIG. 04 By Michael Ezban and David Bayer.
FIG. 05 Courtesy of Clark Erickson.
FIG. 06 Courtesy of Clark Erickson.
FIG. 07 By Michael Ezban.
FIG. 08 By Michael Ezban.
FIG. 09 Courtesy of Clark Erickson.
FIG. 10 Courtesy of Clark Erickson.

Case Study 14
FIG. 01 By and courtesy of Justin Manongdo.
FIG. 02 By Michael Ezban.
FIG. 03 By Michael Ezban.
FIG. 04 By Michael Ezban and David Bayer.
FIG. 05 By and courtesy of Justin Manongdo.
FIG. 06 By and courtesy of Justin Manongdo.
FIG. 07 By Michael Ezban.
FIG. 08 By Michael Ezban.
FIG. 09 By and courtesy of Justin Manongdo.
FIG. 10 By and courtesy of Justin Manongdo.

Case Study 15
FIG. 01 By and courtesy of Justin Manongdo.
FIG. 02 By Michael Ezban.
FIG. 03 By Michael Ezban.
FIG. 04 By Michael Ezban and David Bayer.
FIG. 05 By and courtesy of Sanford Low.
FIG. 06 By and courtesy of Justin Manongdo.
FIG. 07 By Michael Ezban.
FIG. 08 By Michael Ezban.
FIG. 09 By and courtesy of Justin Manongdo.

Part 03
By Michael Ezban.

Representations of Aquaculture Landscapes
FIG. 01 Courtesy of SCAPE.
FIG. 02 Courtesy of SCAPE.
FIG. 03 Courtesy of State Regional Archives in Třeboň.
FIG. 04 Courtesy of State Regional Archives in Třeboň.
FIG. 05 Courtesy of Lateral Office.
FIG. 06 Courtesy of Lateral Office.

FIG. 07 Courtesy of Forbes Lipschitz and Justine Holzman.
FIG. 08 By Michael Ezban.
FIG. 09 By Michael Ezban.
FIG. 10 Courtesy of Smithsonian Library, Washington DC. Image from Victor Coste, *Voyage d'exploration sur le littoral de la France et de l'Italie* (Paris: Imprimerie impériale, 1861).
FIG. 11 Courtesy of James Prosek.
FIG. 12 Courtesy of James Prosek.
FIG. 13 Courtesy of LTL Architects.
FIG. 14 Courtesy of LTL Architects.
FIG. 15 Courtesy of LSU Coastal Sustainability Studio.
FIG. 16 Courtesy of LSU Coastal Sustainability Studio.
FIG. 17 Courtesy of Bibliothèque nationale de France.
FIG. 18 © Illustrated London News Ltd/ Mary Evans. Image from "The Huningue Fish Nurseries," *The Illustrated London News*, February 10, 1861, 193.
FIG. 19 By Michael Ezban.
FIG. 20 By Michael Ezban.
FIG. 21 Courtesy of SCAPE.
FIG. 22 Courtesy of SCAPE.
FIG. 23 Courtesy of SCAPE.
FIG. 24 Courtesy of Estudi Marti Franch.
FIG. 25 Courtesy of Estudi Marti Franch.
FIG. 26 Courtesy of Bibliothèque nationale de France.
FIG. 27 Courtesy of Bibliothèque nationale de France.
FIG. 28 By Michael Ezban.
FIG. 29 Courtesy of Wellcome Collection.
FIG. 30 Courtesy of Wellcome Collection.

Afterword
FIG. 01 Courtesy of David Benjamin.

Index